Smart Cyber Physical Systems

Smart Cyber Physical Systems

Advances, Challenges and Opportunities

Edited by

G.R. Karpagam, B. Vinoth Kumar,
J. Uma Maheswari and Xiao-Zhi Gao

CRC Press is an imprint of the
Taylor & Francis Group, an informa business
A CHAPMAN & HALL BOOK

First edition published 2021
by CRC Press
6000 Broken Sound Parkway NW, Suite 300, Boca Raton, FL 33487-2742

and by CRC Press
2 Park Square, Milton Park, Abingdon, Oxon, OX14 4RN

CRC Press is an imprint of Taylor & Francis Group, LLC

ISBN: 978-0-367-33788-9 (hbk)
ISBN: 978-0-429-32195-5 (ebk)

Typeset in Palatino
by Deanta Global Publishing Services, Chennai, India

Contents

Preface

Smart Cyber Physical Systems (CPS) are co-engineered interacting networks of physical and computational components. A rapidly emerging field, CPS aims to provide the foundation for future smart services, and thus improve our quality of life. CPS is an interdisciplinary field that deals with the deployment of computer-based systems that carry out tasks in the physical world. Identification of the needs, challenges and opportunities in several industrial sectors have accelerated CPS research by identifying and encouraging multidisciplinary collaborative work. The objective of this edited book is to provide a forum for researchers and practitioners to exchange ideas and progress in CPS by highlighting applications, advances and research challenges. A brief introduction about each chapter is as follows.

Chapter 1 addresses the design challenges in CPS.

Chapter 2 has the goal to deep-dive into the SWOT Analysis (strength, weakness, opportunity and threats) of CPS and their applications.

Chapter 3 deals with the various architectures of CPS with different functionalities, and also discusses the different issues and challenges encountered by CPS.

Chapter 4 discusses research on environmental controls like water, ambience, waste management, energy, and safety management systems, and to connect CPS to the creation of intelligent buildings.

Chapter 5 describes the way in which smartness can be incorporated into agriculture, as precision agriculture.

Chapter 6 explains the importance of CPS, highlighting the key physical processes that drive the smart city development and requirements.

Chapter 7 deals with scheduling and resource provisioning, which are considered to be of importance during disaster scenarios.

Chapter 8 discusses the development of a virtual digital twin of the supply chain, which can be used to predict the near-future state of the supply chain.

Chapter 9 provides an overview of war gaming in CPS, its history, design choices, implementation and importance. It aims to simulate the real-time system and provide training to handle all known cyber security vulnerability attacks.

Chapter 10 explores and offers an explanation on how Blockchain is used to enhance a distributed embedded network. Smart Gym has been taken as a case study to witness how Blockchain enabled the CPS to supersede the contemporary.

Chapter 11 introduces the Genetic Sensor API for semantic context-aware computing in CPS. It also discusses the latest technological advances, challenges and research opportunities in realizing context-awareness in CPS.

Chapter 12 deals with social networking and cyber physical social networks. It presents the use of community detection and measures utilized for link prediction in community detection for social Internet of Things.

Chapter 13 outlines some of the top software engineering challenges facing engineering of control systems for Industry 4.0-based manufacturing plants, followed by challenges related to the security of cyber physical production systems.

Chapter 14 introduces the key CPS security trends and potential threats to the modern world. It also covers the key security features for CPS and Industry 4.0 smart factory solutions to deal with the modern threat landscape.

Chapter 15 deals with the web services discovery, which are needed to perform various tasks in CPS. This chapter introduces a Proactive UDDI (pUDDI) framework which will reduce the search space for web service discovery, and a trusted media Blockchain which ensures tamperproof and non-repudiation of data, regarding the quality of the blacklisted web service.

We are grateful to the authors and reviewers for their excellent contributions in making this book possible.

This edited book covers the fundamental concepts and application areas in detail, which is one of the main advantages of this book. We hope that this book, being interdisciplinary, will be useful to a wide variety of readers and will provide useful information to professors, researchers and students

Dr G R Karpagam
Dr B Vinoth Kumar
Dr J Uma Maheswari
Dr Xiao-Zhi Gao

Editors

G.R. Karpagam is a Professor with 24 years of experience in Computer Science and Engineering in PSG College of Technology. She obtained her BE, ME and PhD in Computer Science and Engineering. She is a Senior IEEE member and is operating as the Vice Chair of ACM, local Chapter. She is a recipient of PSG & Sons Teacher of the Year Award. Dr Karpagam has setup several state-of-the-art laboratories, including the Center for Artificial Intelligence Research (AIR), Service Oriented Architecture (SOA) and Cloud Computing, Open Source, funded by industry and funding agencies. She has received research grants from esteemed funding agencies and industrial partners, such as DST, DBT, AICTE, UGC, Cognizant, Cordys, Impiger and Hewlett Packard Enterprise.

Her areas of specialization are database management system (DBMS), data structures and algorithms, SOA, cloud computing, machine learning and Blockchain. Dr Karpagam serves as a reviewer and editor for peer-reviewed national and international conferences and journals. She has published 100+ papers in journals and conferences and is the editor of two books, published by Springer and CRC Press.

B. Vinoth Kumar received the BE degree in Electronics and Communication Engineering from the Periyar University in 2003, and ME and PhD degrees in Computer Science and Engineering from the Anna University in 2009 and 2016, respectively. Dr Vinoth Kumar is an Associate Professor with 16 years of experience at PSG College of Technology. His current research interests include computational intelligence, memetic algorithms and image processing. Dr Vinoth Kumar has co-established an Artificial Intelligence Research (AIR) Laboratory along with Dr G.R. Karpagam at PSG College of Technology. He is a Life Member of the Institution of Engineers India (IEI), International Association of Engineers (IAENG) and Indian Society of Systems for Science and Engineering (ISSE). Dr Vinoth Kumar has published papers in peer-reviewed national and international journals and conferences, and edited a book, published by Springer. He serves as a guest editor/reviewer of many journals, with leading publishers such as Inderscience, De Gruyter and Springer.

J. Uma Maheswari is an Assistant Professor (SG) with 13 years of experience in Computer Science and Engineering in PSG College of Technology. She completed her BE Computer Science and Engineering in Sri Krishna College of Engineering and Technology, Coimbatore from Bharathiar University in 2003 and her ME in Computer Science and Engineering in JJ College of Engineering and Technology, Trichy from Anna University in 2006. Dr Uma Maheswari obtained her PhD Degree in the field of Computer Science and Engineering from Anna University in 2017. Her areas of research include cloud computing and semantic web services. Dr Uma Maheswari has received research grants from esteemed funding agencies like AICTE and DST.

Xiao-Zhi Gao received his BSc and MSc degrees from the Harbin Institute of Technology, China in 1993 and 1996, respectively. He earned a DSc (Tech.) degree from Helsinki University of Technology, Finland in 1999. Since January 2004, he has been working as a Docent at the same university. He is also a guest professor of Beijing Normal University, Harbin Institute of Technology, and Beijing City University, China. Dr Gao has published more than 150 technical papers in refereed journals and international conferences. He is an Associate Editor of the *Journal of Intelligent Automation and Soft Computing* and an editorial board member of the *Journal of Applied Soft Computing, International Journal of Bio-Inspired Computation,* and *Journal of Hybrid Computing Research.* Dr Gao was the General Chair of the 2005 IEEE Mid-Summer Workshop on Soft Computing in Industrial Applications. His current research interests are neural networks, fuzzy logic, evolutionary computing, swarm intelligence and artificial immune systems with their applications in industrial electronics.

List of Contributors

N. Usha Bhanu
Department of ECE
SRM Valliammai Engineering College
Chennai, India

A. Bhuvaneswari
Department of Information Technology
Adhiparasakthi Engineering College
Melmaruvathur, India

Subhrojyoti Roy Chaudhuri
TCS Research & Innovation
Pune, India

I Devi
Department of Computer Science and Engineering
PSG College of Technology
Coimbatore, India

Eshwar. K
Wipro Technologies
Chennai, India

Syed Hameed. M
Caterpillar India Pvt Ltd
Chennai, India

R Rajesh Alias Harinarayan
Department of Computer Science and Engineering
Thiagarajar College of Engineering
Thiruparankundram
Madurai, India

S Hemkiran
Department of Computer Science and Engineering
PSG Institute of Technology and Applied Research
Coimbatore, India

Thanga Jawahar
IoT Tata Consultancy Services
Chennai, India

G R Karpagam
Department of Computer Science and Engineering
PSG College of Technology
Coimbatore, India

M. Keerthivasan
Department of ECE
PSG College of Technology
Coimbatore, India

S. Rithish Kesav
Department of ECE
Coimbatore, India

B. Kishoram
Department of ECE
PSG College of Technology
Coimbatore, India

Amit Kumar
ITC Infotech
Bangalore, India

Sumit Kumar
ITC Infotech
Bangalore, India

J Uma Maheswari
Department of Computer Science and Engineering
PSG College of Technology
Coimbatore, India

M. Maheswari
Department of Electrical & Computer Engineering
K. Ramakrishnan College of Engineering
Tiruchirappalli, India

S. Maheswari
Department of Computer Science and Engineering
National Engineering College
Kovilpatti, India

J. Malligeswaran
Industrial Automation
Abu Dhabi, UAE

Harish Mehra
Industry 4.0
Tata Consultancy Services
Chennai, India

Sujatha Nagarajan
Graxal Games LLP
Tiruchirappalli, India

Swaminathan Natarajan
TCS Research & Innovation
Chennai, India

K Rajkumar
Department of Computer Science and Engineering
Thiagarajar College of Engineering Thiruparankundram
Madurai, India

R. Rekha
Department of Information Technology
PSG College of Technology
Coimbatore, India

R. Roshan
Department of ECE
PSG College of Technology
Coimbatore, India

Dr. G Sudha Sadasivam
Department of Computer Science and Engineering
PSG College of Technology
Coimbatore, India

Dr. S. Mercy Shalinie
Department of Computer Science and Engineering
Thiagarajar College of Engineering Thiruparankundram
Madurai, India

M. Shanmugapriya
Department of Electrical & Computer Engineering
College of Engineering
Anna University
Chennai, India

Suhas Shivanna
Hewlett Packard Enterprise
Bangalore, India

S Sridevi
Department of Computer Science and Engineering
PSG College of Technology
Coimbatore, India

Srikanth Subramanian
Cisco Systems
Bangalore, India

N G Swetha
Department of Computer Science and Engineering
PSG College of Technology
Coimbatore, India

Jitesh Vaishnav
Industry4.0, Tata Consultancy Services
Chennai, India

T. Venkatachalam
Department of ECE
PSG College of Technology
Coimbatore, India

R. Vidhyapriya
Department of Biomedical Engineering
PSG College of Technology
Coimbatore, India

Section 1

My First Journey into the World of Cyber Physical Systems

1

Building Cyber Physical Systems – Design Challenges, Techniques

Usha Bhanu Nageswaran, Maheswari Murali and Shanmugapriya Meiyalagan

CONTENTS

Organization of the Chapter

Section 1 presents the terms and terminologies to help the user to understand the material in this chapter. **Section 2** introduces the concept of cyber physical systems, and **Section 3** describes the physical and logical design of cyber physical systems. **Section 4** shows the adaptive control in cyber physical systems, **Section 5** illustrates the data reliability and security challenges of cyber physical systems, and **Section 6** shows the opportunities and research challenges of cyber physical systems. **Section 7** presents the overview of applications of cyber physical systems in various fields, with **Section 8** depicting the simulation tools/programming frameworks.

1.1 Terms and Terminologies

- Cyber Physical Systems (CPSs): A cyber physical system (CPS) is a mechanism whereby a physical system is controlled or monitored by software algorithms. CPSs are integrated with the Internet and consist of sensors, computing resources and communication capabilities.
- Adaptive Control: In CPSs, adaptive dynamic sensors obtain the real-time values and achieve better parameter optimisation than do static sensors.

- Physical Design: CPSs involve the deployment of physical devices with computational platforms, which are connected to the networking platforms.
- Logical Design: The design involves the end-to-end connectivity between the physical devices and the communication protocols used to design the CPS.
- Security: Cyber security or physical security concepts alone cannot protect CPSs because the integration and communication can introduce vulnerabilities.

1.2 Introduction

A **cyber physical system** (**CPS**) is a mechanism by which physical systems are controlled or monitored by software algorithms. They are integrated with the Internet and consist of sensors, computing resources and communication capabilities. The components of a CPS are broadly classified into a physical process and a cybersystem. The process management is commonly considered to be an embedded system. Embedded systems and networking devices monitor and control the physical processes, which, in turn, affect computations. CPSs involve multidisciplinary approaches, such as digital, analogue, physical and human components integrated together to function as smart intelligent systems. CPSs are considered to be smart systems, such as are involved in smart grid, automated automobile systems, medical monitoring, process control systems, robotics systems, and automatic pilot avionics. CPSs can bring advances in health care, traffic control, fleet management, power generation and delivery, and improve our quality of life in many areas.

1.3 Physical and Logical Design of Cyber-Physical Systems

CPSs require interaction between widespread sensors and actuators in the physical systems, networked to the existing Internet. The optimized design of a CPS plays a major issue in the real implementation of applications to end-user edge devices. The generic representation of a CPS, shown in Figure 1.1, involves the deployment of physical devices in computational platforms, connected well to the networking platforms. The major components of a CPS mainly consist of three design methodologies of Physical, Logical and Network connectivity.

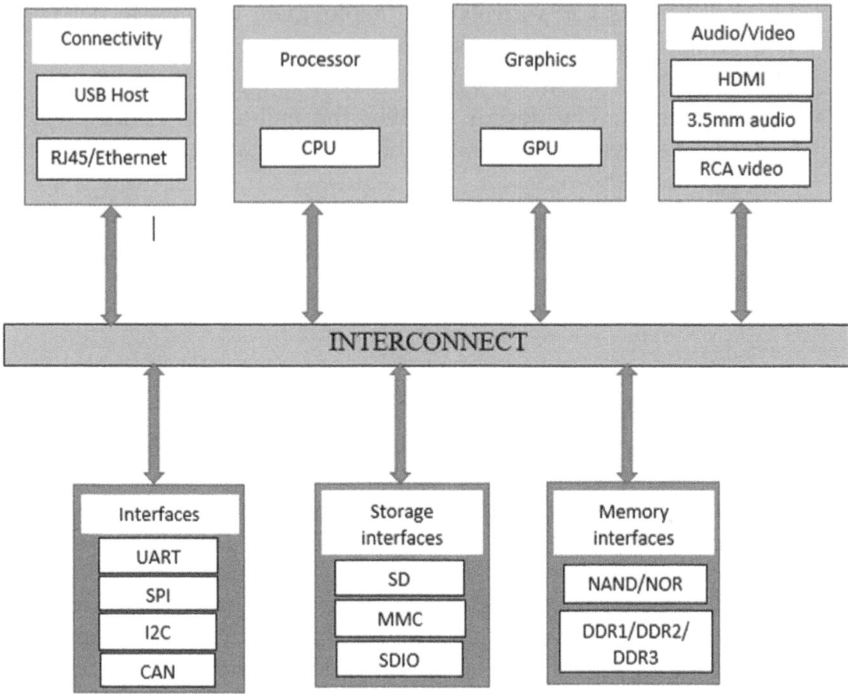

FIGURE 1.1
Generic block diagram of cyber physical systems

1.3.1 Physical Design of a CPS

The physical part is that aspect of the system which is not realised with computers. It includes mechanical parts, and biological or chemical processes, depending on the applications. These include:

(i) Input/output hardware interfaces for sensors and actuators, such as UART, SPI and I2C
(ii) Interfaces for internet connection to the LAN network
(iii) Memory and storage devices, like SD, MMC and SDIO
(iv) Audio/video interfaces, such as HDMI, 3.5 mm Audio or RCA video

Computation Processor

A CPS includes one or more computational platforms, with wireless sensor nodes as inputs and actuators as outputs, and computers with real-time operating systems (RTOS) specific for individual applications.

Network Connectivity

This is the mechanism by which the computer-controlled machines need to communicate, between the computer and the machine, to accomplish a specific task. The integration of the computer with the physical system forms the cyber physical system.

Analysis and Processing

Almost all CPS devices generate data in some form or another, which, when collected and processed by data analytics systems, lead to useful information, to form the basis for taking actions, locally or remotely.

1.3.2 Generic Block Diagram of a Cyber Physical System

The generic block diagram of a CPS [16] consists of a single-board computer (SBC)-based CPS that includes embedded microprocessors, such as CPU, GPU, RAM, storage and various types of interfaces and peripherals.

In Figure 1.1, the widely used single-board mini-computer is a Raspberry Pi processor. The Raspberry Pi processor runs on the Linux operating system and it supports open-source Python programming. Other commercially available processors for Embedded CPS applications include a **PC Duino Development board and Beagle Bone – Black board powered br ARM Cortex processor.**

1.3.3 Logical Design and Protocols for the Implementation of a CPS

The end-to-end connectivity between the physical device and the edge devices determine how the data are sent physically over the network's physical layer or medium. Hosts on the same link exchange data packets over the link layer, using the link layer protocol. The link layer determines how the packets are coded and signalled by the hardware device over the medium to which the host is attached, as shown in Figure 1.2.

The link layer protocols used in the context of the CPS are:

- IEEE 802.3, a collection of wired Ethernet standards for the link layer.
- IEEE 802.11 Wi-Fi, a collection of wireless local area network (WLAN).
- IEEE 802.16 WiMax, a collection of wireless broadband standards, including extensive descriptions of the link layer.
- IEEE 802.15.4, a collection of standards for low-rate wireless personal area networks (LR-WPANs).

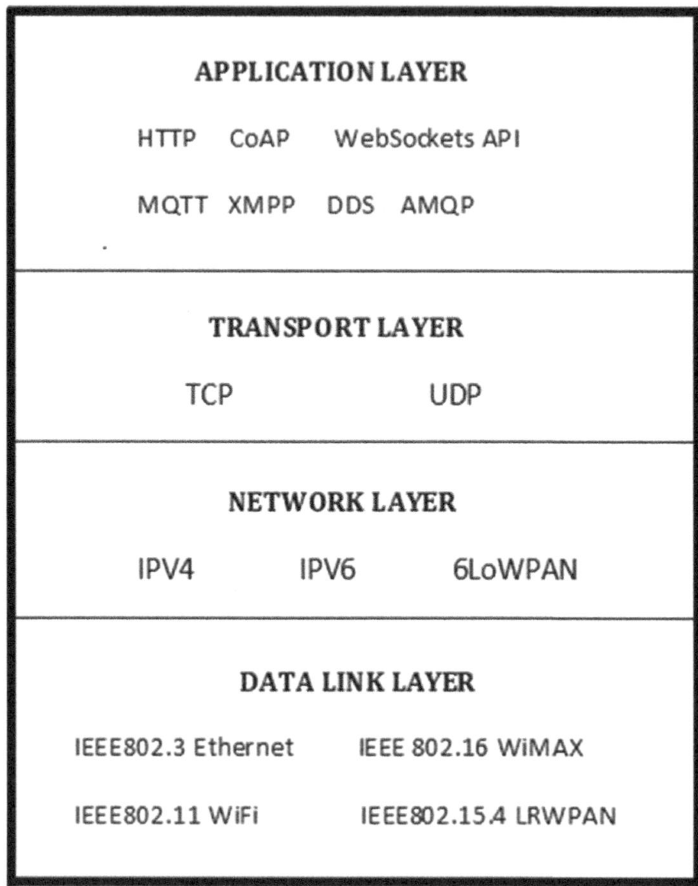

FIGURE 1.2
Logical design of cyber physical systems

1.3.3.1 3G/4G/5G Mobile Communication

The different generations of mobile communication standards include third-generation (3G) (UMTS and CDMA2000) and fourth-generation (4G) (LTE) networks. The CPS systems can communicate over such cellular networks.

1.3.3.2 Network/Internet Layer

The network layer is responsible for sending IP datagrams from the source network to the destination network. Host identification is achieved using hierarchical IP addressing schemes, such as IPv4 and IPv6. The 6 LOWPAN protocol standard is IPv6 over low-power wireless personal area networks. It brings IP protocol to the low-power devices, which have limited processing capability.

1.3.3.3 Transport Layer

Transport layer protocols provide end-to-end message transfer capability independent of the underlying network. The connection-oriented transmission control protocol (TCP) and the connectionless user datagram protocol are used.

1.3.3.4 Application Layer

Application layer protocols define how the applications interface with the lower layer protocols to send the data over the network. Application layer protocols enable process-to-process connections, using ports.

1.3.3.5 Hypertext Transfer Protocol (HTTP)

The Hypertext Transfer Protocol (HTTP) is the application layer protocol that forms the foundation of the World Wide Web (WWW). The protocol is based on a client–server model, where the client sends a request to a server using the HTTP commands. HTTP is a stateless protocol and each HTTP request is independent of the other requests.

1.3.3.6 Constrained Application Protocol (CoAP)

The Constrained Application Protocol (CoAP) is an application layer protocol for Machine-to-Machine (M2M) applications

1.3.3.7 WebSocket

The WebSocket protocol allows full duplex communication over a single socket for sending messages between client and server.

1.3.3.8 Message Queue Telemetry Transport (MQTT)

Message Queue Telemetry Transport is a lightweight messaging protocol based on the client, where a number of smart CPSs connect to the server, known as a MQTT broker, and publish messages on topics on the server. Extensible Messaging and Presence Protocol (XMPP) is used for real-time streaming and communication of XML entities.

1.3.3.9 Data Distribution Service (DDS)

Data Distribution Service is a data-centric middleware standard for device-to-device or machine-to-machine communication.

1.3.3.10 Advanced Messaging Queuing Protocol (AMQP)

Advanced Messaging Queuing Protocol is an open application layer protocol for business messaging.

1.3.3.11 REST-based Communication Application Programming Interfaces (APIs)

Representation State Transfer (REST) is a set of architectural properties by which one can design web services and web APIs, that focus on a system's resources, and determine how resource states are addressed and transferred. The client–server, stateless, cacheable, layered system, uniform interface and Code-on-Demand are the REST architectural constraints.

1.3.3.12 WebSocket-Based Communication APIs

WebSocket APIs allow bi-directional, full duplex communication between client and servers. They do not require a new connection to be set up for each message to be sent.

Modelling of the CPS system is shown in Figure 1.3. The system consists of multinodes placed in different locations for monitoring physical parameters. The edge devices/nodes are equipped with various sensors. The end nodes send the data to the cloud in real time, using the WebSocket service. The data are stored in the cloud database. The data analytics are carried out in cloud centres to aggregate the data and make predictions.

1.4 Adaptive Control in Cyber Physical Systems

In general, a CPS consists of embedded control processors (ECPs) that take outputs from the sensors and process the sensor values to control the actuators, to achieve the desired results as shown in Figure 1.4.

The performance of the ECPs is decided, based on the software as well as the hardware that are used to control the applications. Today's embedded processor design is based on the parameters received from the sensors. All the sensors' parameters are taken from the field for a fixed-time interval that is decided for the worst-case situation, based on the validity of the system. When these fixed-design systems are applied to applications that require fewer of the parameters to be measured, it becomes inefficient.

For example, when the control system designed to measure parameters, such as temperature and pressure, for a short time period from a system which has high bandwidth, is used for another system, which has a shorter bandwidth and requires the measurements of fewer parameters, then its resources are not being properly used; in addition, it consumes more energy. This makes the whole system inefficient [1]. In order to improve the performance of the system, the sensor parameters should be optimized. For better performance of the system, adaptive dynamic sensor parameter optimization has proved to be better than static sensor parameter optimization.

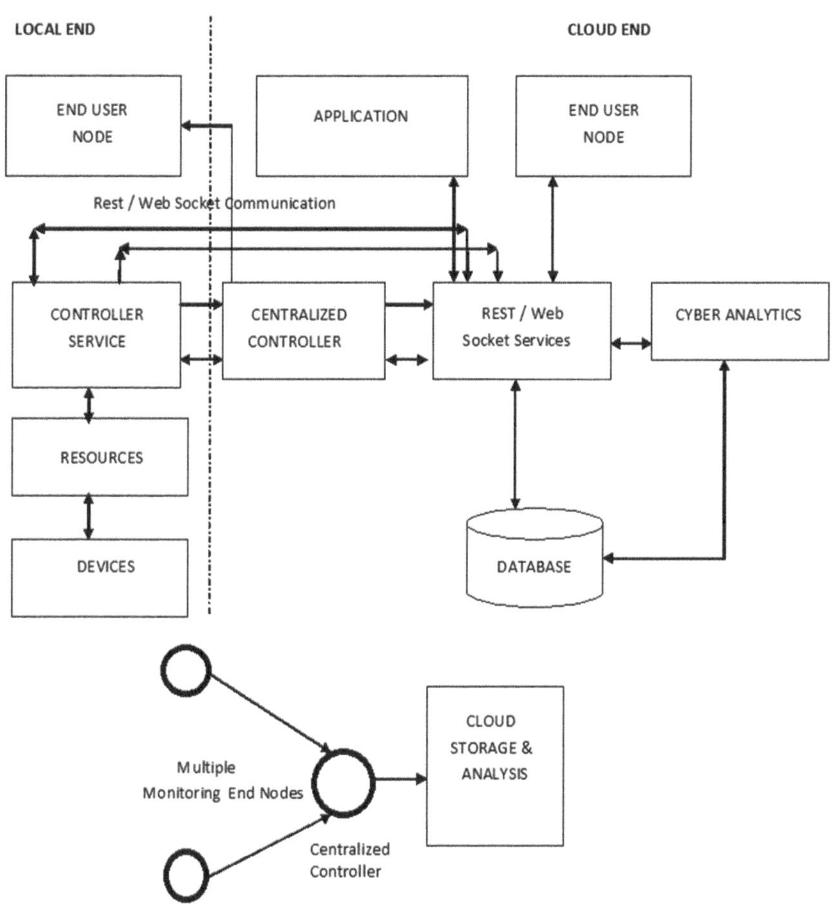

FIGURE 1.3
Modelling of generic cyber physical systems

Figure 1.4 shows adaptive control in a CPS which optimizes measured signals from the sensors.

Several optimization techniques have been presented for sensor parameters, in both adaptive and static modes. In [1], the authors applied Reinforcement Learning (RL) to optimize the sensor parameters. This resulted in a reduction in average power consumption of the embedded controller and, in turn, for the overall CPS. The use of RL for optimization does not require previous knowledge about the system. RL learns the environment from its experience and does not need any training to be given. This makes the optimization of the control part easy. Traditional time-varying control theory and the multiple application-specific model have been used to optimise the parameters [2]. A scheduling algorithm for a shared computation platform was developed [3]

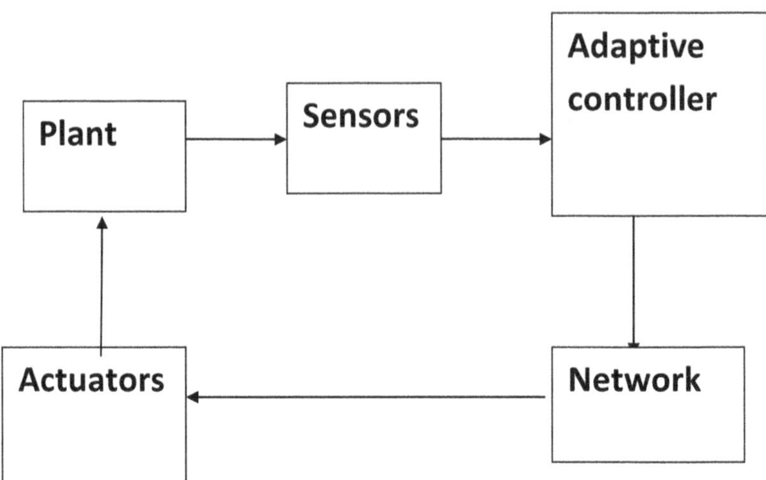

FIGURE 1.4
Adaptive control in cyber physical systems

by the same authors, that obtains feedback from a physically embedded controller to provide better control applications. This algorithm requires a complex scheduling strategy. In [4], the authors proved that the adaptive method of assigning a sampling period for sensor measurement online is better than the static time period assignment for a sampling time. In [5], for a heterogeneous cyber physical system which uses different sensors, various sampling schemes have been proposed to achieve better results.

1.5 Security and Privacy Issues in a Cyber Physical System

In CPSs, the physical process is monitored or controlled for the given application. As the interaction between the physical and cyber systems increases, the security vulnerabilities in the cyber system also increase. Cyber security or physical security concepts alone cannot protect CPS because the integration and communication between them can introduce vulnerabilities. Physical attacks may damage or compromise availability of information on the device, and cyberattacks can cause physical malfunctions. Security and privacy must be key considerations for CPS design, development and operation.

1.5.1 Definition – Security and Privacy

Security is a set of measures to protect the information and resources of the system with respect to confidentiality and integrity, and to accomplish its intended goal. Privacy is an aspect of security often thought of as freedom from observation, eavesdropping and disclosure of confidential information. Confidentiality means that information is not released to unauthorised persons. The terms 'confidentiality', 'integrity' and 'availability' are interconnected, to provide the system's security.

- Confidentiality – information is kept secret between the authorised parties.
- Integrity – assets can be modified only by authorised parties or only in authorised ways.
- Availability – assets are accessible to authorised parties at appropriate times.

1.5.2 Security and Privacy Issues in CPS

In normal situations, the CPS applications are independent. In cases of emergency, these applications need to interact with each other and to share their resources to carry out emergency tasks. Traditional security solutions are not suitable for heterogeneous CPS applications. One of the most important concerns of CPSs is to protect sensitive or private data, to guarantee user privacy and ensure the system's trustworthiness. Security and privacy attacks on CPSs may be due to the interconnection between devices, interconnection with the internet and because of the deliberate actions of malicious users or unauthorized third parties. The vulnerable places for CPS security are shown in Figure 1.5.

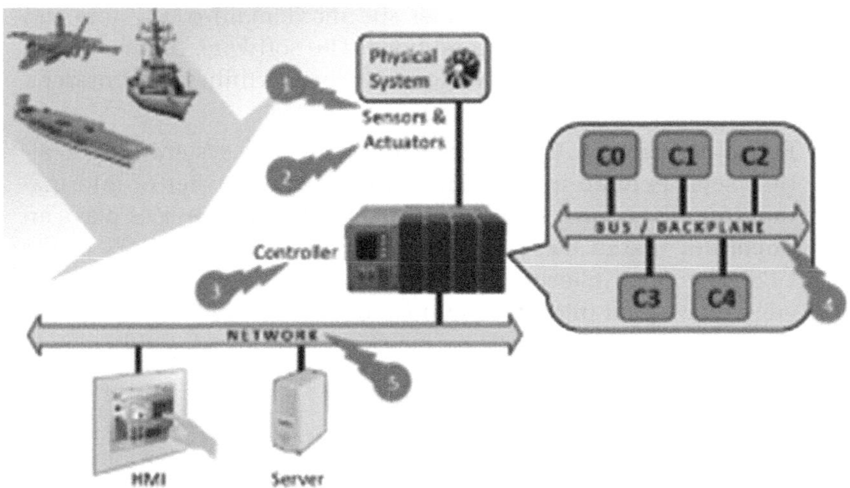

FIGURE 1.5
Attacks on cyber physical systems

1.5.3 Security Objectives in a CPS

i) Confidentiality is the ability to prevent information and data from being exposed to any unauthorised individual or party from inside or outside the system. Confidentiality is achieved by applying encryption algorithms on the data stored in the system and transmitted in the network, and by restricting access to the places where the data appear. In a CPS, confidentiality is ensured by protecting communication channels from eavesdropping, to prevent the system status from being hacked.

ii) Integrity is the ability to keep data as they are and to prevent any unauthorised manipulation. In a CPS, integrity is ensured by preventing all possible ways of attack and imposing security measures to protect the CPS's physical goals.

iii) Availability is the ability of all subsystems to work properly and have their work completed on time and when needed. Availability ensures that all CPS subsystems are functioning correctly by preventing hardware, software and network failures, power failures and denial-of-service (DoS) attacks.

iv) Authenticity is the ability to guarantee the veracity of all parties participating in any CPS processes. Authenticity must be achieved in all subsystems and processes to have an authentic and genuine CPS.

v) Robustness is the degree to which a CPS can continue to work properly, even in the presence of limited disturbances. It is closely related to reliability, which can be provided by fault tolerance.

vi) Trustworthiness is the degree to which people can rely on the CPS to perform required tasks under specific domain constraints and according to specific time conditions. The software, hardware and collected data must all show a level of trustworthiness to consider a CPS to be feasible and trustworthy.

vii) Reliability. Since a CPS consists of a physical system with real-time sensors in an internet environment, the transfer of information from the source sensor elements to the edge devices plays an important role in measuring the Quality of Service (QoS) of the CPS. Therefore, filtering algorithms need to be integrated in the edge gateway for filtering out unwanted disturbances from the sensor inputs to the CPS.

1.5.4 Threats Against Cyber Physical Systems

- Attempt to perpetrate a denial-of-service attack
- Attack against open ports and services

- Attempt to change device settings
- Attempt to change control settings
- Attempt to generate Man-in-the-Middle attacks on data exchange involving the CPS

1.5.5 Security Challenges to CPSs

Security in CPSs must include protection of both cyber and physical aspects. The potential cyber-attacks, with physical consequences, must be predicted and mitigated. The challenges to CPSs are listed below.

- **Physical security**: A CPS consists of a variety of devices such as sensors, actuators, controllers and interconnection devices. If the CPS is working in isolation, security is not taken into consideration in the design of most CPSs. The physical security has been almost the only security measure employed to protect the resources from attack.
- **The real-timeliness nature**: Real-time decisions in CPSs are crucial for the interactions between physical and cyber aspects. Accessing the timeliness assists in designing better risk-assessment, attack-detection, and attack-resilient solutions. If cryptographic mechanisms are enforced to provide secure communication, it could cause delays that could affect some real-time deadlines. Therefore, lightweight hardware-based mechanisms should be considered.
- **Insider threat**: Insiders could create security problems, either intentionally or unintentionally. Unintentional cases occur where insiders unknowingly use an infected laptop or USB stick, that could give remote attackers access points to the CPS. The insider threat has been underestimated and overlooked, and it certainly needs serious considerations

1.5.6 Security Solutions

- **Network-level security solutions**: Major concerns in CPS communications include keeping the data private and allowing only authorised access. Network attacks can be implemented at the physical layer as well as the software layer. Cryptographic mechanisms can protect the data from network threats. Intrusion detection and authentication mechanisms can limit any adverse effects.
- **Device-level security solutions**: Each device can be identified using a unique label stored in permanent storage. The device identity cannot be changed. Authentication mechanisms should be applied to limit the serious problems associated with identity theft.

1.6. Design Challenges and Opportunities of Cyber Physical Systems

The CPS consists of interactions between cyber systems and the physical world. It involves interactions between heterogeneous systems, consisting of distributed computing systems and sensors and actuators in a real-time environment. The objective of this section is to provide an overview of future challenges and opportunities in the field of CPSs for engineering research applications. The advances in the field of science and technology play a vital role in deploying the application of CPSs in the real world. The implementation of CPSs also takes care of the economic and societal needs of the end-user requirements.

1.6.1 Opportunities and Challenges in the Applications of CPSs

The innovative design of the physical elements of CPSs, with actuators and sensors, needs to be integrated with communication networks to implement a wide range of application. The opportunities for the field in the next five to ten years are enormous, with technological advances and customized end-products. The implementation of CPSs should be specific to the economic and societal needs of the nation.

The challenges and opportunities in the design of CPSs require perfect synchronization with sensors and actuators. The devices should be capable of handling large volumes of data and their location characteristics. The typical research direction for the next-generation CPSs involves the following parameters:

- At the physical end, the devices, sensors, actuators and the network elements need to be integrated to handle large volumes of data.
- The parameter at the virtual end requires changes in the parameters of end-to-end protocols for connectivity, address scheme in the MAC layer and requirement of energy conservation between large number of connected nodes.

In spite of developments in the design of smart systems, with heterogeneous computation and interoperable networks, the characteristics and performance of CPS can be enhanced by the following parameters:

(i) **Need for Unified Design, Modelling and Architecture**

The smart physical systems, with a distributed computing environment, may have the capability to dynamically adapt with context changing based on the operating conditions of the sensed environment and the user's context for the given application. Therefore, a unified architecture has to be realized to make CPSs in reality, and to self-adapt to the field under investigation.

(ii) **Self-Configuring Systems**

In real-time implementation of a CPS, it should have the self-configuring capability to allow large numbers of physical devices to be integrated to accomplish certain functionality. The CPSs may all have self-configuring capability, allowing a large number of physical devices to be integrated in the cyber world. Therefore, in real time, the CPS must be adaptive for modelling in the specific applications, and be resilient to failures of individual sensors/actuators in the physical device and networked components.

(iii) **Interoperable Communication Protocols**

The CPS needs to support a large number of interoperable communication protocols in the physical link layer, network, transport and in application layers. This enables the physical systems to communicate with other devices and networking components.

(iv) **Unique Addressing, Verification and Validation**

The CPSs consist of heterogeneous hardware and software, and the whole physical system needs to be identified with unique identifiers. The CPS has intelligent interfaces which adapt, based on the context modelling between networked users and the real-time environment. The CPS also requires end users to query the physical devices, monitor their status and need to be controlled by remote with the management infrastructure. This feature allows the CPS to achieve high degrees of interoperability, dependability and reconfigurability.

(v) **Latency and Performance**

For delay-sensitive applications of a CPS in the field of vehicular, aerospace, surveillance, gaming and network systems, the end-user satisfaction and performance depend on the latency in the accomplishment of the inputs to sensors and outputs from actuators. The perfect modelling, well-defined abstraction level, with highly efficient tools, needs to be proposed to achieve its implementation.

(vi) **Security and Scalability**

In critical secured applications, such as medical health care systems, interconnected networks and smart grids for electrical power systems, the transformation of information and the reliability of the data are very important. Self-healing architectures may be proposed for a secured CPS to avoid systems failure and cyberattacks. Since a CPS has to be deployed to handle a wide range of applications, the physical devices and the networking node need to be scalable for future enhancements in the static and dynamic environments.

(vii) **Low Power Consumption**

In the design of future CPSs, it is necessary to have energy efficiency in terms of power optimization for integrating a wide range of physical devices with the cyber data. This ensures low-cost implementation for the end user.

1.7 Overview of CPS Applications

The cyber physical system integrates networked computers, called embedded controllers, with the physical world, through networked sensors. It is found in a wide range of applications, such as agriculture, medicine, industrial manufacturing, smart grid, transportation, education, environmental monitoring, and smart city, etc. In this section, an expanded overview of the applications of CPS is described.

1.7.1 Agriculture

Agriculture is an important sector of the Indian economy. The GDP of any country depends heavily on agriculture. In addition, most of the raw materials of many industries, such as sugar cane, cotton, grains, jute, etc., come from agriculture. Hence, it is important to develop the agriculture sector by adopting and adapting new technologies. The implementation of CPSs in agriculture will surely increase the production and will also minimise energy consumption. CPSs have been applied for smart agricultural field monitoring to control pests [6], in particular, the development of a 'rat detection system', using advanced pest control design solutions. This method of CPS application to agriculture significantly improved crop yield and hence achieved higher incomes for farmers. Many research papers have been published to enrich precision agriculture using CPSs. A smart potato monitoring system, for example, has been developed using CPSs, adapting precision agricultural management to ensure high yield [7]. A smart green agricultural system can also be adapted, using CPSs to monitor the growth of the plant from seed to its marketing. The complete process can be automated without input from the farmer.

1.7.2 Medicine

CPSs have a very important role to play in the health care industry. Many high-accuracy sensors have been used to continuously monitor the patient's health. Also, the information can be transmitted to the doctor concerned through the Internet of things (IoT) under high security [8]. In addition to continuous health monitoring, it is possible to carry out operations remotely. ACPS-based robot, to assist home patients, has also been implemented.

1.7.3 Industrial Manufacturing

The role of CPSs in industry will improve production and also the safety of the people working. In industry, various sensors that control the operation of the machines communicate through the network. The embedded controller designed for industrial manufacturing is based on a microprocessor and a high-end microcontroller, which receives inputs from various sensors that are connected to the machines. The inputs are processes in embedded controllers and the corresponding actuators will be controlled. A 5G wireless network that connects the sensors will achieve high-speed operation of the process.

1.7.4 Smart Grid

In the 21st century, electricity is the essential requirement, replacing fossil fuels. Hence, the demand for electricity has increased dramatically. It is easy to transmit electricity from one place to another through grids. Nowadays, smart grids are replacing conventional grids which interconnect various physical grids, with smart controllers regulating the transmission of electrical energy. CPSs come into the picture in smart grids as a smart embedded controller with interconnection to physical grids through a communications network. A CPS provides dynamic interconnection between physical grids and provides timely responses for better transmission of energy. In [9], the authors explored the application of CPSs to smart grids and its research challenges.

1.7.5 Transportation

The role for CPSs in transportation is very high. In transportation, CPSs can be used to monitor the traffic by using intelligent traffic-measuring sensors in different geographical locations. The information is shared to all the users so that they can avoid the high-traffic routes. Also, the interest in unmanned vehicles is increasing significantly, and sensing, communication and control over the unmanned vehicle is being designed and implemented using CPSs. In [10], the authors counted the number of vehicles travelling from one location to another, whereas in [11], Wireless Sensor Networks and CPSs have been explored for use with unmanned vehicles by analysing sampling patterns, communication, scheduling and resource sharing. Vehicular CPSs has also been investigated to achieve negligible delay, even for high-speed users [12].

1.7.6 Education

Nowadays, many advanced technological methods are being used for education in this country. Many open-platform teaching methods, using CPSs, have been adopted in education which makes the learning process simpler. Cloud-based teaching can also be adopted using CPSs. The virtual classroom is being used by many institutions to enhance the teaching experience. In [13], the authors explored multi-level classroom implementation, using CPSs.

1.7.7 Environmental Monitoring

Environmental monitoring, using advanced technology, is in great demand in places such as forests, high mountains and rivers where humans cannot enter and stay for long periods, to measure particular environmental parameters. Wireless sensor networks play an important role in monitoring the environments of such places. High-speed wireless networks with highly accurate sensors are needed to measure environmental parameters in such places. CPSs can be exploited to monitor the surrounding parameters, using sensors, and the data are communicated to the control unit immediately, so that immediate action can be taken. If needed, an actuator can also be activated to take immediate action and to rectify the effects of poor environmental conditions in these remote places. In [14], the authors investigated the role of CPSs in monitoring the environment during flooding and storms.

1.7.8 Smart City

CPSs offer enormous opportunities for the operation of the smart city. 'Smart city' implies an urban area with innovative technologies, to be implemented in the field of transportation, environmental monitoring, healthcare, ambience, social services, energy distribution, uninterrupted water supply, smart drainage system, smart leak identification process for water supply, etc. For all these implementations, CPSs can play an important role. Various sensors are placed for different purposes, and the information collected from the sensors is communicated to the embedded control unit through a high-speed communications network. In [15], the authors investigated the use of CPSs in implementing a smart city, and also listed research issues for the use of CPSs with interconnecting heterogeneous networks.

1.8 Simulation Languages/Tools Used for CPSs

The deployment of CPSs involves hardware, software and networking solutions suited to specific applications.

- The model-based design and development of CPSs can be implemented using tools, like Simulink and LabVIEW, which form the basis for simulation of embedded control systems.

- Programming language models for the simulation of embedded real-time systems can be carried out by tools like PTIDES (Programming Temporarily Integrated Distributed Embedded Systems).
- For a real-time CPS, the Giotto time-triggered language for embedded programming can be used.
- For modelling and simulation of heterogeneous embedded systems, the Ptolemy programming language can be used for real-time CPSs.
- Tools, like the PESSOA programming language, can be used for synthesising controllers in CPSs.
- For designing CPSs, using Macro Lab, the programme can be divided into subprogrammes for loading the respective nodes with sensors and network elements.
- Cloud storage models and API, using Web Application Messaging Protocol (WAMP)-AutoBahn for IoT solutions, can be used.
- Xively Cloud for IoT focuses on the front-end infrastructure and devices for smart applications.
- Python Web Application Framework Django can work with different databases for data analytics.
- Amazon Web services for IoT solutions, like EC2, Auto scaling, S3, RDS, Dynamo DB, Kinesis and SQS, can be used, using Python programming

The objective of this chapter is to explore the design challenges in CPSs, such as selecting suitable embedded processors and their interfacing with the sensors, the selection of low-cost sensors for different applications, and for efficient control and monitoring systems. In the near future, CPSs will revolutionise how humans can interact with the physical system around us, controlling and monitoring the various elements.

References

1. Buinil, Hamid Mirzaei 2017. Adaptive embedded control of cyber-physical systems using reinforcement learning. *IET Cyber-Physical Systems: Theory & Applications* 3: 1–9.
2. Sala, A. 2005. Computer control under time-varying sampling period: An LMI gridding approach. *Journal of Automatica* 41(12): 2077–2082.
3. Henriksson, D. and A. Cervin 2005. Optimal online sampling period assignment for real-time control tasks based on plant state information. *IEEE Conference on Decision and Control, CDC-ECC'05*, 4469–4474.
4. Cervin, A., et al. 2011. Optimal online sampling period assignment: Theory and experiments. *IEEE Transaction of Control System Technology* 19(4): 902–910.

5. Albertos, P. 1997. Real time control of non-uniformly sampled systems. *Control Engineering Practice* 7(4): 445–458.
6. Ciprian-Radu, et al. 2014. Architecture model in the field of precision agriculture. *International Conference on Agriculture for Life, Life for Agriculture*, 196–200.
7. Mehdipour, F. 2014. Smart Field monitoring: An application of cyber-physical systems in agriculture. *International Conference on Advanced Applied Informatics (IIAIAAI)*, 181–184.
8. Lee, I. et al. 2012. Challenges and research directions in medical cyber-physical systems. *Proceedings of the IEEE* 100(1): 75–90.
9. Xinghuo, Yu and Yusheng Xue 2016. Smart grids: A cyber physical system perspective. *Proceedings of the IEEE* 104(5): 301–306.
10. Zhou, Y., et al. 2016. Privacy preserving transportation track measurement in intelligent cyber-physical road systems. *IEEE Transactions on Vehicular Technology* 65(5): 3749–3759.
11. Wan, J., et al. 2013. From machine-to-machine communications towards cyber-physical systems. *Computer Science & Information Systems* 10(3): 1105–1128.
12. Kumar, N. et al. 2015. Optimized clustering for data dissemination using stochastic coalition game in vehicular cyber-physical systems. *Journal of Supercomputing* 71(9): 3258–3287.
13. Lawlor, O., et al. 2015. AERO-beam: An open-architecture test-bed for research and education in cyber-physical systems. *Industrial Electronics Society, IECON 2015-Annual Conference of the IEEE*, 5080–5086.
14. Sierla, S., et al. 2013. Common cause failure analysis of cyber-physical systems situated in constructed environments. *Research in Engineering Design* 24(4): 375–394.
15. Cassandras, C.G. 2016. Smart cities as cyber-physical social systems. *Engineering* 2(2): 218–219.
16. Bahga, Arshdeep and Vijay Madisetti 2015. *Internet of Things: A Hands On Approach*. Universities Press.

2

SWOT Analysis of Cyber Physical Systems (CPS)

Srikanth Subramanian

CONTENTS

Organization of the Chapter

Section 1 presents the terms and terminologies for the user to understand the chapter.

Section 2 introduces the SWOT framework for a cyber physical system, which provides a high-level summary of the subsequent chapters.

Section 3 highlights the key and unique strengths of a CPS.

Section 4 examines in detail the weakness or challenges associated with CPS. The challenges are classified into three main domains – physical, technical and security challenges.

In Section 5, the opportunities for CPS to make a difference in different verticals – transportation, manufacturing and healthcare, to name but a few, are discussed. Also, the social and environmental impacts of CPS are discussed in this section.

Section 6 focuses on the main threats to CPS. The threats are mainly analyzed from the security perspective, namely what is the nature of the threats and what is the impact they can potentially have on the systems.

Network and social threats that pose a challenge to CPS are also discussed.

Section 7 is forward looking, discussing how CPS can play a significant role in transforming the legacy of modern systems built today by capitalizing on the technological advances, to efficiently cater for the demands of the future. Several case studies, in the areas of manufacturing and healthcare, are discussed with a forward-looking outlook.

Section 8 provides an insight into how data and analytics is a fulcrum for CPS to continuously learn and adapt, enabling it to be a predictable and accurate system.

2.1 Terms and Terminologies

CPS Cyber Physical Systems

SWOT Strengths, Weakness, Opportunities and Threats

SOA "Service-Oriented Architecture, a style of software design where services are provided to the other components by application components, through a communication protocol over a network."[1]

Augmented Reality "A technology that superimposes a computer-generated image on a user's view of the real world, thus providing a composite view."[1]

IoT "Internet of Things, a system of interrelated computing devices, mechanical and digital machines, objects, animals or people that are provided with unique identifiers and the ability to transfer data over a network, without requiring human-to-human or human-to-computer interaction."[1]

2.2 SWOT Analysis of Cyber Physical Systems (CPS)

The relevance of CPS is unquestionable and unparalleled in today's fast-paced world, powered by technology in almost every sector.

The cyber physical system, with its ability to network cyber systems (computation and communication) and physical systems (sensors and actuators), has the potential to significantly impact our daily lives through the digitization of smart services. The cyber physical systems have multiple applications in diverse, real-time and complex fields, like healthcare, air-traffic control systems, ATMs, power systems, and automated transport systems, among others. It is becoming increasingly difficult to identify systems which are not cyber physical, given the penetration of electronics and software into virtually all facets of our lives. The application of CPS varies from a smaller deployment, like an Internet Protocol (IP) camera or a home router, to a huge complex deployment, such as a power grid. Moreover, the concept is inherently multi-disciplinary and multi-technological, and relevant across vastly different domains, with multiple socio-technical applications.

With any advances in technologies, where human intelligence drives the creation of machines that are super intelligent and which exceed human capabilities, there is always a balance between advantages and disadvantages.

This chapter aims at carrying out a SWOT analysis (Strength/Advantages, Weakness/Challenges, Opportunities and Threats) of CPS (Figure 2.1).

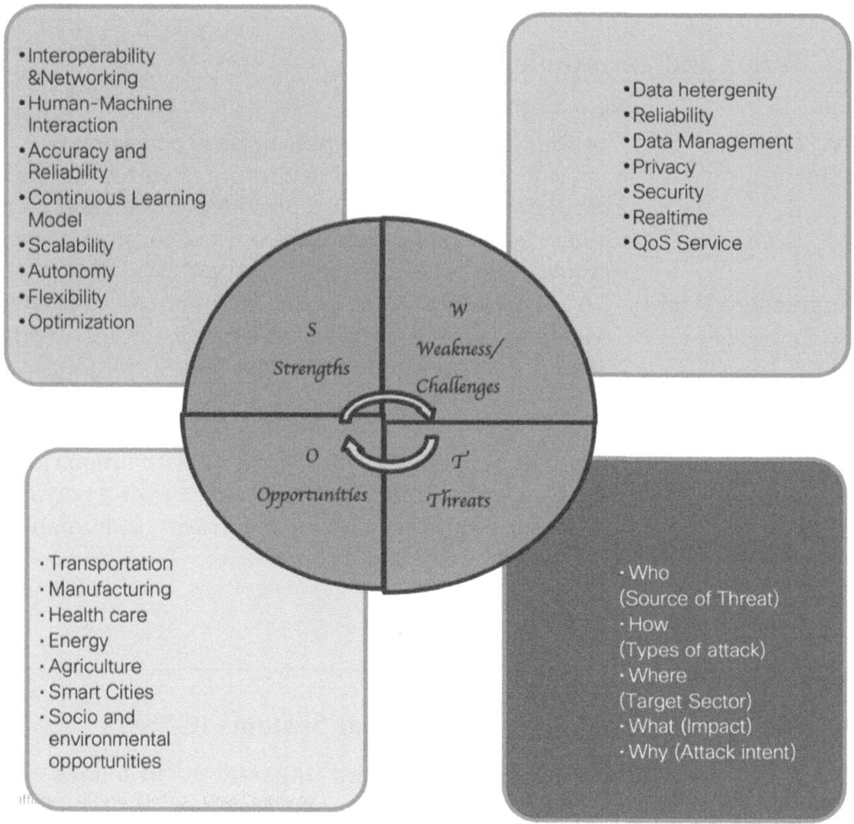

FIGURE 2.1
SWOT analysis of CPS

2.3 Strengths/Advantages of CPS

CPS is a promising solution for the integration of the physical and cyber worlds, because of several advantages it offers.

2.3.1 Interoperability and Networking

CPS has the inherent advantage of its interoperability capability with software-defined networking (SDN) in both hosted and cloud platforms. Hence, this makes it easier to deploy CPS in any enterprise-level architecture of any scale.

2.3.2 Human-Machine Interaction

Human form a core and integral part of CPS, so that CPS is built with a core design model, based on human and machine interactions, helping to design a simple and comprehensive system.

2.3.3 Dealing with Certainty

The greatest strength of CPS is its ability to deal with uncertainty and to perform well in an unreliable ecosystem.

The framework on which the CPS is built provides this ability to deal with such uncertainty and unreliability.

2.3.4 Accuracy

With advances in metrics systems, the ability to measure data with utmost precision has soared beyond the nanoscale to the pico and femto scales. The CPS taps into this advance very well and provides highly accurate information.

2.3.5 Continuous Learning

Advances in machine learning and comprehensive algorithms help train the CPSs continuously, through continuous feedback, making it a highly reliable system overtime.

2.3.6 Better System Performance

Built on a highly sophisticated framework of hardware and software computational resources, with reliable monitoring and alerting systems in place, the CPS offers a very high system performance.

2.3.7 Scalability

CPS combines the power of the physical and cyber domains to design large-scale systems and to meet the demands of cloud computing. The physical domain involves a combination of human, mechanical motion control, sensors and the chemical/biological processes driving them. The cyber domains involve software modelled with complex algorithms, built by state-of-the-art programming tools on a highly expandable network infrastructure.

2.3.8 Autonomy

CPS is highly autonomous, due to the complex sensors integrated with the cloud. There is a feedback mechanism that drives auto-learning and -correction, thus enabling an adaptive system.

2.3.9 Faster Response Time

The faster response time is largely attributed to the power of the computational resources. This helps in early detection and pre-emption of failures, resulting in a robust system

2.4 Weaknesses/Challenges

While there is an ultimate goal of using cyber physical systems to transform the ecosystem into a digital and smart world, there are multiple hurdles that need to be overcome in the process, due to the numerous constraints being posed.

2.4.1 Physical Challenges

2.4.1.1 Bridging the Cyber and Physical Worlds

The physical environment is very difficult to model, due to various moving elements, so that developing a cyber interface with it is very cumbersome.

There needs to be a continuously evolving training model that is fed back into the system to train the systems.

2.4.1.2 Defined Boundaries in the Developing World

This is closely associated with the previous problem, in that there is no boundary to the control elements, so it is very difficult to describe a model as being complete when it has to be continuously evolving and being trained.

2.4.1.3 Networking Stability in Predictable Complex Systems

Heavy reliance on the network stability poses a challenge for the cyber physical system. Though robust network designs and outlays are created, there is still latency and interworking issues, that make the cyber physical systems unreliable at times.

2.4.1.4 Bandwidth

To model a reliable CPS, there is a great dependency on infrastructure bandwidth. The more complex the deployment becomes, the greater are the bandwidth requirements, so that a virtual dependency is established with the bandwidth providers.

2.4.1.5 Technical Challenges

There are a number of technical issues that challenge the effectiveness of CPS.

Data Heterogeneity. Data heterogeneity sometimes results in huge variance in data samples, leading to inaccurate results in CPS and causing a failure. The variance could be due to actual differences in observations in real-life modeling, thus making it difficult to model a complex real-life system.

Reliability. The reliability stems from the unpredictability of the environment that is being modeled, so a continuous evolution is required. CPS is expected to continue to work reliably under unexpected circumstances and to adapt itself in cases of failures through continuous learning and feedback mechanisms.

Data Analytics. Data storage, aggregation and analysis from different sources of data are very challenging. CPS is expected to model a predictable behavior, based on the analysis of data feed, both real-time and historical data, from multiple incompatible and complex data systems. A robust big-data system and highly efficient machine-learning algorithms are required to power-up an efficient cyber physical system. With continuous advances in technology today, this becomes very challenging.

Privacy. With a large volume of data, especially sensitive information, being handled by CPS as they are deployed in some verticals where anonymity and privacy are of paramount importance, as in the banking sector and government agencies, a comprehensive privacy and data protection policy is needed to ensure privacy of the data being managed by CPS. Due to a substantial rise in the number of hackers, data theft always poses a huge challenge in such deployments.

2.4.1.6 Security Challenges

Security vulnerabilities pose an enormous challenge for CPS. The vulnerabilities can enter into the system from both the hardware and software elements, and potentially cause minor to critical damage to the data/system. Thus, CPS has the challenge to tackle the security threats to both the hardware and software resources, to ensure a threat-free system.

The hardware, software and network mesh opens up a vulnerable medium for threat exposure and allows an attack. The complexity of attack can vary, ranging from impacting a smaller deployment of a CPS, such as routers, IP cameras, mobile phones etc., resulting in individual losses, to effects on more complicated scenarios, like power grids, sensitive government agencies and financial institutes, that affect a large number of people. There have been many infamous attacks, like the Mirai malware attack, which impacted several websites including Twitter, Netflix, Github etc., and data thefts from famous social media websites, that have happened in the recent past. The hacking community is continuously finding more innovative and intelligent ways to hack into the systems.

There is a lot of research work underway in several top institutes and research labs to protect the CPS from such attacks and to preserve the integrity and authenticity of the systems.

2.5 Opportunities in Cyber Physical Systems

With the continuous evolution of technology and the way data/information are used as a fulcrum for model systems, there is a great opportunity for CPS to make disruptive changes in multiple verticals. It also opens up opportunities in new markets.

CPS is a game changer in many fields and promotes innovation. This has provided opportunities in different verticals to grow the markets significantly.

A combination of both emerging and well-established technologies, such as Modular Access Systems (MAS) for modular designs, Service-Oriented Architectures (SOA) (to provide interoperability in heterogeneous systems), Cloud Computing (for high-speed processing), Big Data (to achieve coherence, patterns and analytics from data stored in high volumes), Machine-to-Machine (M2M) (to enable the interconnection among devices and continuous learning) and Augmented Reality (to support the integration of human feedback), have all provided a framework for advances in CPS, enabling the development of a powerful model. Such revolutions have significantly helped in increasing the implementation of large-scale systems, improving the adaptability, autonomy, efficiency, functionality, reliability, safety and usability of CPS.

The following sectors are expected to benefit significantly through technology advances in CPS.

- **Transportation**. In the automotive industry, CPS technologies promise to greatly reduce the annual death toll from accidents, caused by human error, through early detection and prevention of accidents by smart sensors deployed in the automobiles. CPS also promises to play a significant role in an effective traffic management system. Some airspaces in United States are already piloting the use of CPS for air safety and the avoidance of air traffic congestion. The large amounts of data stored also enables the creation of several smart applications for passenger navigation, flight status and estimated time of arrival (ETA), terminal information, etc.

- **Manufacturing**. CPS is of great value to the manufacturing industries, to improve productivity, reduce cost and waste and increase operational efficiency. CPS, through automated sensors and actuators, operates the machines and proactively signals any faults in the system. CPS has a huge role to play in transforming the industries to Industry 4.0.
- **Healthcare**. CPS, due to its high level of accuracy, has great potential to be used as a complex and accurate measurement and control element in healthcare. The artificial cardiac pacemaker, a small device that is placed in the chest or abdomen to help control abnormal heart rhythms, is a proven example of advances in this field. The healthcare domain needs a significant amount of interaction with the human/physical world, and CPS provides that flexibility through both digital and physical data acquisition.
- **Energy**. CPS has the potential to play a significant role in developing an efficient energy management system for the consumers, through automated sensors and switches that help in balancing the renewable energy requirements. Through integration with solar power energy systems, it promises to offer a highly cost-effective solution, and this is a huge, largely untapped market that is expected to grow significantly in the next decade.
- **Agriculture**. With global warming resulting in unpredictable climatic factors and changes in seasonal behaviour, agriculture is heavily impacted. One way to combat this is through the use of technology. The CPS technologies could increase sustainability and efficiency (less waste) through exploiting the value chain. Integration of technology with agriculture has become imperative. Intelligent robots, temperature and moisture sensors and aerial images will further drive the agricultural sector to be more profitable and efficient.

2.5.1 Social Impact of CPS

Smart Cities. Opportunities for CPS in smart cities abound, due to the present situation. By powering a digitally connected world, and leveraging the data framework, CPS helps in solving many common problems faced today by any big city in every country, such as traffic congestion, pollution and waste management, and parking issues, and enables the modeling of smart connected communities, lighting, operation systems, safety and security systems.

CPS, with an open software-driven foundation laid on a platform, with innovations powered by analytics, automation and architecture as the backbone, is driving enormous advances in various sectors.

2.5.2 Environmental Impact of CPS

CPS could play a major role in addressing two of the greatest challenges of our time, namely global warming coupled with shortage of sustainable energy supplies.

There is great potential to tap the energy coming from the ground (geothermal energy) to cut down the energy consumption of two major consumers – transportation and buildings. Use of renewable sources of energy, such as solar, through a tighter interconnect between the cyber and physical worlds, will significantly help in reducing the energy consumption of commercial and residential buildings. This is a major reason for today's corporates to invest heavily in renewable sources of energy like solar and wind turbines.

Likewise with the advances of CPS, the design of transport systems offers highly scalable and energy-efficient modes of transport.

2.6 Threats in Cyber Physical System

Today's digital world is fully interconnected, with numerous electronic devices, platforms and infrastructures, which represent the starting point of a threat. The threat is largely driven by the communication medium and the vulnerabilities associated with it. There are various levels of threats for CPS starts, ranging from the destruction of a small device to disruption of a single enterprise to damage to a complete infrastructure. The target ecosystems vary from personal information theft to a national impact, with threats and endangerment to lives.

The hardware and software mesh of cyber physical systems, that have open-source software and cloud-based infrastructure, provide easy access to the attackers to gain entry into the system and cause damage or destruction. The impact can vary, based on the nature and severity of the attack, ranging from financial implications to loss of life. Hence, a strong secure firewall is critical for a CPS to operate in a closed environment with tight security boundaries.

It is important to have an efficient and continuous risk assessment, monitoring and resilient delivery system in real time to safeguard from threats. A large CPS is an interconnect between several hardware and software resources on the same communication network. So, when designing a good security system, it is not just the hardware and software that should be borne in mind, as the network must also be taken into account too, as easy access to the network can destroy the entire chain of CPS. Though the networks are robust, infrastructures are highly scalable and deployments are complex, the hacking population is growing and the advances in technology are making it easier to breach the defences of a network, rendering the CPS more vulnerable to attacks. It is always important for CPS technology to keep pace with the advances in hacking technology, to build appropriate security measures.

2.6.1 Taxonomy of Threats

Below is presented the taxonomy of threats, sourced from the literature and real-life experience.

The taxonomy is based on the five basic principles related to the threats:

- **WHO** are the threats (**Sources of Threat**)?
- **HOW** do the threats happen (**Types of Attacks**)?
- **WHERE** are the threats (**Target Sector**)?
- **WHAT** are the threats (**Impact of Attack and Incident Categories**)?
- **WHY** do the threats occur (**Attack Intent**)? (Figure 2.2)

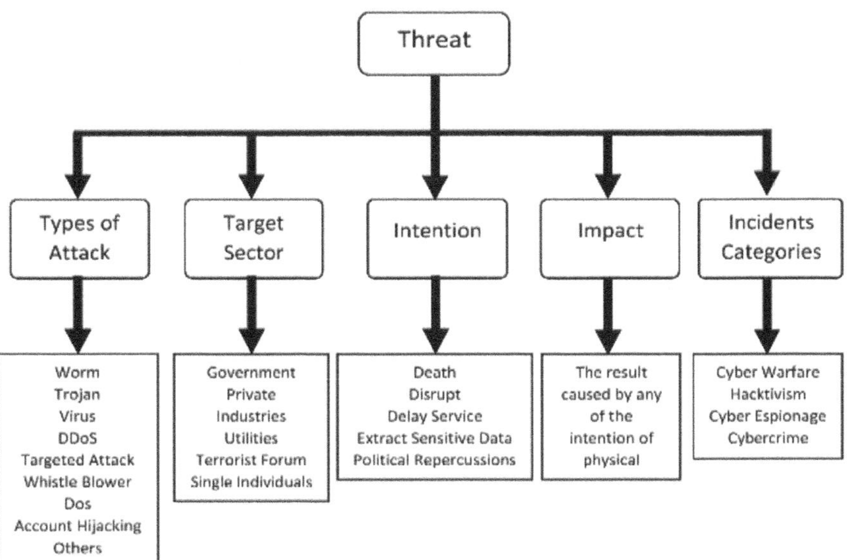

FIGURE 2.2
Threat taxonomy

2.6.1.1 Sources of Threat

The originator of threats is, in most cases, a human. The various classes of originator, based on real-life evidence, are:

- Cyber criminals.
- Disgruntled employees.
- Terrorists, activists and organized criminal groups.
- Computer specialists trained in hacking.
- White-collar employees with malicious intent.
- Nation states, governments, due to political feuds between states and countries.

2.6.1.2 Types of Attack

There are various ways by which an attacker can attack a system. Some of the commonly used attacking mechanism prevalent in cyber space are virus attacks, Distributed Denial of Service (DDoS), botnets, hacking, malware, pharming, ransomware, spam etc.

2.6.1.3 Targeted Sectors

The targeted sector for a security threat ranges from a single individual to as large as an organization or a government sector. Typically, government organizations, banking and industries are more prone to cyber-attacks due to the impact of the attack.

2.6.1.4 Intention

The most common intentions are theft, espionage and the compromising of national security.

2.6.1.5 Impact

Based on the research literature, five areas of impact, as a result of a security attack have been identified.

- Physical/digital.
- Economic.
- Psychological.
- Reputational.
- Social/societal.

2.6.2 Network Threats

Of the sophisticated networks operating today, 85% use encrypted HTTP/HTTPS traffic. It allows flow of data to be more secure, so that the communication channel preserves its integrity and the sensitivity of the data. However, there is a hidden issue with data encryption. If bad data are encrypted, there is no way to distinguish them from the encrypted good data, and hence they can be transported in the network without detection.

One way to detect malicious software variants ("malware") in a network is by a technique called "Traffic Finger-Printing." This requires continuous monitoring of network traffic to watch for a pattern of malicious packets being transported in the network. However, the hacker community has grown smart enough to enable them to bypass these security measures, by injecting additional dummy packets along with the malicious ones, randomizing the patterns and establishing new patterns, so that it becomes a huge challenge to implement a robust detection mechanism.

Much work has gone into securing the transport layer and internetwork of systems, through the implementation of SSL (Secure Socket Layer) and TLS (Transport Layer Security), to help in building a robust and secure cyber physical system.

There are a few key things, in terms of security threats, to be aware of in order to take an informed approach to designing a cyber physical system.

- It is a question of "when," rather than "if," the security perimeters will become compromised, so planning for prevention/mitigation of a security breach is absolutely critical when designing a CPS.
- Attacks are becoming more frequent, complex and sophisticated, with new sources emerging regularly.
- Existing software/hardware/network infrastructures, tools, processes and practices need to be leveraged to help companies protect their assets.
- Multiple technology companies and research organizations have made significant investments and have generated considerable momentum in the security space.

2.6.3 Social Threats Due to Cyber Physical Systems

- Possible Terminator-type scenarios
- Unemployment
- Loss of purpose in life
- Over-reliance on computers, with machines gaining self-awareness

2.7 Future Advances in Cyber Physical Systems

There are multiple fields where CPS has the potential to make further huge impacts. Two areas – Industry and Medical – have been taken here as references and the growth potential and impact of CPS in these two areas will be analyzed here.

2.7.1 Industry 4.0 – Future Smart Industries

Industry 4.0 is the greatest revolution powered by cyber physical systems. CPS has enabled the evolution of manufacturing sectors to focus on highly efficient production at a lower operational cost. The advances in hardware, software systems and powerful analytical systems, with a combination of real-time/historical data and fast and accurate cloud computing mechanisms, are the reasons behind this transformation (Figure 2.3).

The technological advances in cyber physical systems have enabled a major paradigm shift in the *modus operandi* of today's factory *versus* 4.0 factories. The focus on today's operational model is based on automation of actuators and fault detection systems, through application of smart sensors.

The goal of the factory operations is to operate with a lean workforce management, through automated machineries and reduced wastage. It is a big leap from earlier functional models, where the focus was on "output generation" and not on the process or its efficiency.

However, with the advances in CPS, considerable innovation has been inputted to further "smartify" the industry.

The goal transitions from "quick remediation" to "prevention" and "uptime" of production units.

		Today's Factory		Industry 4.0 Factory	
	Data source	Key attributes	Key technologies	Key attributes	Key technologies
Component	Sensor	Precision	Smart Sensors and Fault Detection	Self-Aware Self-Predict	Degradation Monitoring & Remaining Useful Life Prediction
Machine	Controller	Producibility & Performance (Quality and throughput)	Condition-based Monitoring & Diagnostics	Self-Aware Self-Predict Self-Compare	Up Time with Predictive Health Monitoring
Production System	Networked manufacturing system	Productivity & OEE	Lean Operations: Work and Waste Reduction	Self-Configure Self-Maintain Self-Organize	Worry-free Productivity

FIGURE 2.3
A comparison of today's factory and 4.0 factory operational model

Continuous alerting and monitoring services, combined with a powerful 6C system, that consists of Connection (sensor and networks), Cloud (computing and data on demand), Cyber (model and memory), Content/Context (meaning and correlation), Community (sharing and collaboration) and Customization (personalization and value), can enhance the information system and help in detecting machine degradations, wear-and-tear and wastage in production, thus enabling a smart-connect-production System. This is the evolution of the Factory 4.0 era.

2.7.2 Proactive Healthcare

Nothing can be more significant than improving the health and well-being of humans and increasing their quality of life. Thus, the role and advances of CPS in the healthcare domain is extremely critical.

In healthcare environments, the acquisition of data retrieved from sources such as:

- Real-time monitoring and computational systems.
- Passive monitoring and computational systems.
- Computerized investigation records of patients.

will aid in analysis of the patient diagnosis analysis.

The decision making, based on these data, has still not been fully explored and exploited and is a current focus of active research by various organizations, universities and research labs. Several startups have been actively funded by governments and angel investors to accelerate such research.

The studies involve a combination of body area networks, implantable smart devices, that can be a substitute for a body part, and programmable monitoring devices for computation, control and action.

The CPS in healthcare applications can be divided into two areas: (a) assisted and (b) controlled.

2.7.2.1 Assisted

Health monitoring for "assistance," without any interference to an individual's normal living, falls under this category. Vital body function data, acquired from assisted devices through bio-sensors in real time, can be used to provide sound medical advice to the patient. This helps in treating multiple out-patients and elderly people who are being proactively monitored.

2.7.2.2 Controlled

A controlled environment is an environment (intensive care unit (ICU) or a high-dependency unit (HDU)) fully under intense clinical supervision all the time. In hospitals, the data from multiple automated medical devices and manual checkups is collated to provide an accurate diagnosis of patients. The advances in CPS should enable the development of a feedback system, using all these clinical observations, into an intelligent decision-making system, providing an auto-preventive 100% reliable remedy/cure mechanism (with minimal involvement from humans) for patients.

Sequence of events depicted in Figure 2.4:

(1) Data are collected from patients and the sensor data are transmitted *via* the gateway to cloud storage.

(2) Sensor data are sent to the cloud server and analysed with powerful computers, where the data are processed in real time.

(3) A comparative analysis of the patient's historic data is made with the data in the cloud storage.

(4) The central observation centre is notified about the availability of a new patient record for investigation.

(5) Clinicians from the observation centre investigate the patient's records processed in the cloud server.

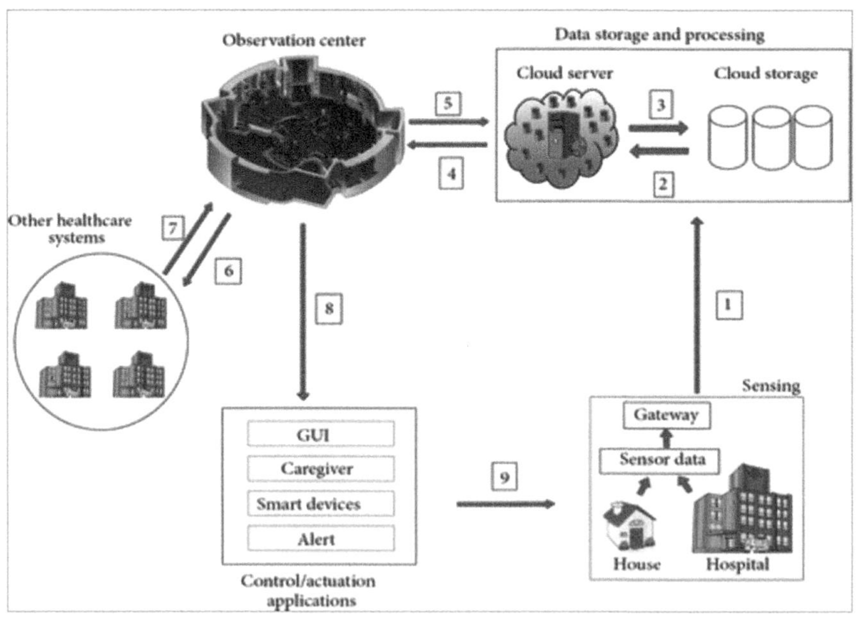

FIGURE 2.4
A CPS for healthcare based on published research

(6) Any further necessary clinical validation is carried out, with other distributed, specialized healthcare systems, to determine the clinical accuracy with respect to specific parameters.
(7) Information is processed and validated from other distributed, specialized healthcare systems
(8) The final investigation report on the patient is transmitted to the control/actuation application
(9) The patient is provided with appropriate medical guidance, based on the investigation reports.

As a further advance on this, human intervention can be further reduced by making CPS even smarter. For instance, with a patient in ICU, with a pattern of fluctuating blood pressure readings, the CPS monitoring service can detect this pattern and provide auto-recovery to the patient by triggering an intravenous (IV) line to regulate it.

It can be imagined how accurate and reliable the CPS must be to function in this case, as it deals with a life. This represents a highly ambitious goal to reach, but it would be appropriate when CPS realizes its full potential. With the powerful combination of humans making informed decisions with the help of smart CPS, the life and quality of humans can be substantially improved under both clinical surveillance (Controlled Environment) and at home (Assisted Environment).

2.8 Role of Data Integration and Analytics

With the growth of cyber physical systems, data integration, analytics and monitoring become key elements for the success of deployments.

Big data plays a game-changing role and provides an edge to many businesses today, which can make informed decisions based on the internal and external data, thereby achieving a major positive impact on productivity and service. Many manufacturing systems lack smart analytics tools. The Industry 4.0 revolution, based on cyber physical system, is driving the manufacturing world towards building such technologies and innovations. Germany is leading the world in driving this revolution. CPS brings in a mesh of hardware and software resources, running powerful algorithms, that help achieve strong predictive analysis and enable operational efficiencies in many industries.

For powerful predictive analysis, data becomes a major source. Data becomes the fulcrum of the CPS ecosystem. The different machines and systems connect very well, so that data feeds from all machines are used

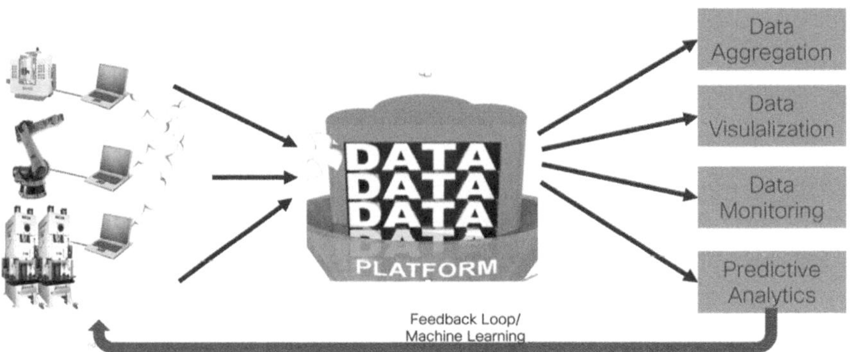

FIGURE 2.5
Data framework for continuous learning

accurately for the analysis and in making intelligent decisions. Productivity and particularly production cost effectiveness increase, due to such powerful data monitoring and analytics (Figure 2.5).

Data Aggregation. The data platform/pipeline is defined to aggregate data from different control elements/machines, logically grouped and formatted for consumption at various levels.

Data Visualization. Creative visualization mechanisms (such as dynamic charts) are created for consistent historical and real-time reporting, using the data aggregated. Some critical actions are taken, based on the data visualized at regular intervals.

Data Monitoring. The real-time data from physical systems are continually monitored, with appropriate threshold settings and appropriate alerting mechanisms in place. This ensures that the system always operates under a controlled environment, with preventive care being taken to avoid any problems.

Predictive Analytics. Based on the aggregated data, a predictive algorithm runs to predict the next set of events, thereby enabling the system users/operators/business owners to optimize the operational model to suit the predictive data.

There is a continuous learning process involved with this and hence the feedback loop is critical to such data to the system, which continuously learns and adapts. With a large volume of learning and adaptation, the output from the systems becomes very accurate, making it very efficient and reliable.

2.9 Conclusion

Technology today is built over distributed computing and crowdsourcing of information. The power of collective intelligence is a key success of CPS. Our ecosystem is being driven by this digital evolution and transforming it to develop smart, connected cities through the advent of such technologies. CPS requires a highly skilled workforce and research work promoting collaborations between industries and universities. The capability of such systems manifests itself due to the combined power of human intelligence and machine power.

The evolution of technology today, like IoT and CPS, has enormous potential to take the world by storm, change our lives significantly and encourage the present and future generations to embark on a new journey to a next-generation digitally connected smart world.

Bibliography

Online Sources

1. https://en.wikipedia.org/.
2. http://cyberphysicalsystems.blogspot.com/2011/07/cyber-physical-systems-advantages-and.html.
3. https://www.nap.edu.
4. https://journals.sagepub.com.
5. http://www.infosecurity-magazine.com.
6. http://cyberphysicalsystems.blogspot.com/.

Electronic Journals

7. Martin Miskuf, Iveta Zolotova. "Comparison between multi-class classifiers and deep learning with focus on Industry 4.0", *2016 Cybernetics & Informatics (K&I)*, 2016. ISBN Number: 978-1-5090-1834-5
8. Fei Tao, Qinglin Qi, Lihui Wang, A.Y.C. Nee. "Digital twins and cyber–physical systems toward smart manufacturing and industry 4.0: Correlation and comparison", *Engineering*, 5(4). (2019), 653–66.
9. Yosef Ashibani, Qusay H. Mahmoud. "Cyber physical systems security: Analysis, challenges and solutions", *Computers & Security*, 68. (2017), 81–97.

Section 2

Smart Cyber Physical Systems – User Cases

3

Building Cyber Physical Systems

Context: Wireless Sensor Networks

Vidhyapriya Ranganathan and Rekha Rajagopalan

CONTENTS

Organization of the Chapter

Section 1 presents the terms and terminologies used in the manuscript to help the understanding of readers. **Section 2** introduces cyber physical systems (CPS), followed by the motivation towards the development of CPS. **Section 3** discusses the network formation and architecture of CPS. **Section 4** describes the applications and challenges encountered by CPS. **Section 5** provides an analysis of some of the design requirements of CPS and presents some of the major technical problems associated with CPS. **Section 6** gives a summary of this chapter as well as discussing directions for future work.

3.1 Terms and Terminologies

Smart systems, Data-to-information conversion, Cognition, Cyber space, Localization, Synchronization, Resource management, Sensor deployment, Mobility, Data aggregation

3.2 Introduction

The word 'cyber' indicates the combination of computers, computer networks, network security and technologies related to computer control. Man-made and natural systems that are governed by physics and which operate in continuous time are termed physical systems. Hence, CPS is an interdisciplinary field that combines the components that are physical in nature with the virtual world, such as information technology. Embedded computers and networks are used by CPS to integrate analysis of sensor-collected data, transfer of information from one place to another and the control of physical elements in a system.

A CPS is considered to be like a Wireless Sensor Network (WSN), with a data mining module added to obtain new information from the sensed data. WSNs are networks of small, low-powered, wireless devices that have sensing, communication and on-board processing capability. WSNs are specific to a particular field, whereas CPS do not have such restrictions.

This chapter shows the architecture of a CPS and discusses the technical and research issues handled by CPSs.

3.2.1 Integration and Motivation

One of the important challenges faced by CPS is the heterogeneity of physical systems and their interactions. The heterogeneity can be explained with a sample scenario. Consider a patient with a chronic disease, such as heart disease, who is in need of continuous health status monitoring by healthcare providers. Firstly, the patient may be residing at his smart home, which is a CPS, with security cameras, light bulbs and other electronic devices connected to the home automation system. Secondly, he may be wearing a watch that has computing capability (CPS) and is capable of measuring the speed at which his heart contracts. A temperature sensor may also be present to monitor the external temperature. The collected information needs to be remotely monitored by a healthcare provider and there should also be a feedback to the patient, based on decisions made on the basis of the continuous monitoring. Each of the CPS described above may use different communication protocols and control mechanisms, and hence it may not be possible to connect them together in an easy manner to make these devices work together. A solution is therefore required to achieve interoperability in such situations. A component-based technology was suggested by [1] to overcome issues related to interoperability.

In recent years, research has focussed largely on building CPSs that are capable of tolerating inaccuracies in the sensed data and of still providing good quality of service and performance. Nowadays, CPS incorporate more complex integrations with many complex systems for building smart systems in the field of healthcare, energy, manufacturing and transportation. Cyber threats and security attacks highlight the importance of designing appropriate security and privacy policies for CPS.

3.3 Network Formation and Architecture

3.3.1 Types of Architecture

CPS have a range of applications in various areas, such as industry, healthcare, military and transport. Some of the design requirements of CPS are presented below.

3.3.1.1 ISA–95 Architecture

ISA–95 architecture [2] proposed by the International Society of Automation (ISA), divides CPS into five levels, and is shown in Figure 3.1. Level 0 includes the physical production process in smart factories. Levels 1 and 2 provide manufacturing control, such as sensing, manipulating, monitoring,

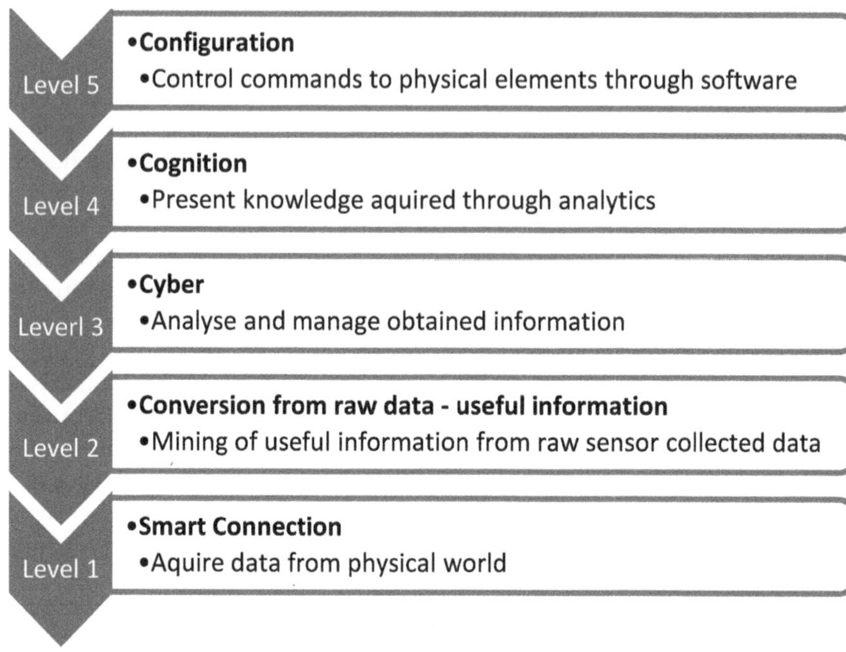

FIGURE 3.1
ISA – 95 Architecture of CPS

supervisory control, and automated control of the production process. Level 3 deals with producing the desired end products and optimizing the production process. Level 4 is concerned with business planning, production scheduling and operations management.

3.3.1.2 5C Architecture

A level 5 architecture for manufacturing application was proposed by [3]. The stages of the architecture include smart connection, data-to-information conversion, cyber, cognition and configuration. Selection of the appropriate sensors and consideration of the different types of data from the various machines comes under Level 0 of the architecture. Conversion of the collected data to meaningful information is the focus of Level 1.

Specific data analytics techniques will be applied to the information collated to achieve analysis of performance of a particular machine or to make a prediction of future behaviour of a particular machine, based on currently gathered information and historical information. Presentation of analytics results to users and their contribution to decision making form the cognitive level of the architecture. The feedback from the cyber space to the physical space, applying the corrective or preventive decision taken at the cognition level to the monitored system, represents the configuration stage.

Ahmed et al. [4] proposed an architecture with five main modules, namely (1) sensing module: to sense data from the physical world; (2) data management module: to perform heterogeneous processing of the sensed data; (3) next-generation Internet: to present applications with a choice of paths between the data management module and the service aware module; (4) service-aware module: to perform decision making, task analysis and task scheduling, using the sensed data, and to send them to the services available; and (5) application module: available services are assigned to different applications in this module.

3.3.2 8C Architecture

An 8C Architecture was proposed by [5] for smart factories. This was obtained by adding three more levels to 5C Architecture. The additional levels to the existing 5C Architecture were (i) Coalition – this module focuses on integration of the value chain and the production chain between different parties in terms of the production process; (ii) Customer – this module focuses on customer participation in the design and development of a product. Customers can give their suggestions, track the progress of the designed product or can even make modifications in the product specification during the production progress; and (iii) Content – this module focuses on maintaining product traceability record, such as product-developed environment and also after-sales service record.

3.3.3 Medium Access Control (MAC)

MAC protocols focus on avoiding collisions while accessing a network. Cyber physical systems include devices of various computational ranges and interconnection among different types of network. Hence, for handling devices and networks of different configurations by CPS networks, the MAC protocols need to be adaptive in nature.

- Virtual Back-off Algorithm: [6] proposed a structure that uses a sequence number, K, which is set, based on characteristics of different networks. A Learning Automaton (LA) is placed at every node in the network and base station, and this LA determines the K value best-suited for that network. After finding the optimal K values throughout the CPS, the throughput is calculated and compared with the threshold throughput. If the calculated value exceeds the threshold, then the selected combination of K value set is rewarded; otherwise, it is penalized.

3.3.3.1 Routing

- Real-time routing using nRTEDBs: [7] developed network-enabled real-time embedded databases (nRTEDBs) and integrated them to wireless sensors to provide real-time services in CPS. To achieve routing immediately when the event occurs, several nRTEDBs are deployed in the sensing area which periodically broadcast beacon messages so that other sensors in the vicinity can know the shortest path to reach the nRTEDB. Load balancing between nRTEDBs is performed by increasing the beacon period, so that paths from sensors to that particular nRTEDB will expire earlier, preventing an nRTEDB from becoming overloaded. When an nRTEDB detects an event, it communicates with the nRTEDBs that can be approached in a single hop. It increases the communication scope after exchanging a certain number of event messages.

- Genetic algorithm-assisted routing: [8] developed a routing algorithm to enhance the network lifetime through mobile agents. In order to find the paths that connect every source node available in the network, Dijkstra's shortest path algorithm is used. The optimal path to communicate the sensor-collected data to the base station is identified with the help of a genetic algorithm.
- Routing based on hybrid systems: The algorithm proposed by [9] includes selection of the best routing method offline and its determination online.

3.3.3.2 Localization

The objective of localization is to detect and track objects by finding their physical coordinates. Some of the commonly used localization techniques are discussed below:

- RFID-based localization: Radio Frequency Identification (RFID) localization has been widely applied to localize objects due to its small size, low cost and lack of a need for a power supply. The RFID-based techniques rely upon received signal strength (RSS) between reader and tag to obtain location information. Environmental factors influence the accuracy of RFID-based localization techniques. Phase information was used as an indicator by [10] to estimate the difference between the tag and the various antennas. In order to find the location of the tag, a hyperbolic positioning method is used. An orientation-aware phase model was developed by [11] to deal with orientation-based problems in RFID localization schemes.
- Localization based on Global Positioning System (GPS): Vehicular cyber physical systems make use of GPS for localization. The advantage of this scheme lies in its simplicity. GPS-based schemes assume the availability of non-disruptive signals from at least two satellites with sufficient quality. Due to high-rise buildings and flyovers, there is considerable degradation in performance in GPS-based

localization schemes. [12] developed geometry-based localization for vehicular CPS. The GPS outage problem was solved by using knowledge of the changing aspects of vehicles and the curves and lines present on the roads.

3.3.3.3 ***Synchronization***

CPS need to achieve high levels of coordination among the physical and computational systems in order to obtain a dynamic response to the system demands. The requirements to achieve synchronization are discussed below:

- Clock Synchronization: Each node in a CPS has its own clock and there may be a natural drift in a clock over time. Clock synchronization is established and maintained by commonly used protocols such as Network Time Protocol (NTP), Precision Time Protocol (PTP) and Global Navigation Satellite Systems (GNSS).
- Timing requirement within a node: In order to handle the time delay during the input-output process and proper scheduling in CPS, the input read by the sensors, the output written and the computations performed should be completed on time. Certain architectures are capable of identifying and storing the actual time at which a particular event was sensed and will also be able to perform actuation at appropriate time points [13].
- Timing requirement within a network: Latency must be bounded and small in order to achieve overall CPS performance. Time-sensitive traffic may be generated by one CPS node and received by one or more CPS nodes. Hence, one of the various CPS nodes can be designated to be the CPS Network Manager (CNM). The principal task of the CNM is to collect time-sensitive stream requirements from all the available CPS nodes and communicate that collected information to the network. Moreover, when the network is ready to communicate the time-sensitive data stream, CNM transfers this information to the CPS nodes.

3.3.3.4 ***Security***

The various security requirements of CPS are:

- Device access security: Physical access to CPS nodes needs to be secured. Unauthorized access to devices may lead to modification and manipulation of the system.
- Data transmission security: Data transmission over CPS networks may be disrupted by Denial of Service attacks by attackers. CPS that

aim to reduce the operational cost rely mostly on open networking standards, which are more vulnerable to attack during data transmission.

- Data storage security: CPS nodes are mostly small and lightweight. Complex cryptographic mechanisms cannot be used to protect the data sensed and stored by such nodes. Hence, lightweight security mechanisms should be used to protect the stored data.
- Actuation security: The top level in the CPS architecture is to issue control commands from a computer to the physical elements involved in the physical system. Based on the commands received by the physical elements, an appropriate actuator action is initiated. Protection is required to ensure that this feedback command is received from a trusted source.

The various attacks on CPS are:

- Delay attack: The attack is carried out by delaying any of the clock synchronization messages. Due to delayed/incorrect timing messages, the calculations made by the node, using the incorrect information, will also be wrong. [14] conducted a study to analyse whether a CPS has been attacked or not.
- Attacks at data collection layer: Tampering with the CPS nodes, physical destruction of devices, or compromised encryption keys, which target the confidentiality, availability and integrity of the entire CPS, are examples of attacks at this layer.
- Attacks at the transmission layer: Intercepting any transmitted data in the network ('eavesdropping'), pretending to be an authorized user and thus gaining access to the transmitted information, followed by modification, deletion, insertion of new information to the existing data ('spoofing') or 'flooding' are some of the common attacks in this layer.
- Attacks at the application layer: Common attacks at this layer include unauthorized access to user devices, user data leakage or injecting malicious codes into user applications.

3.4 Moving from WSN to CPS

3.4.1 Applications

- Intelligent transportation: Smart transportation CPS focus on reducing road accidents, fuel consumption and congestion, and on improving transportation safety. Vehicular *ad hoc* network (VANET) is used by CPS for communication between vehicles. [15] investigated how parameters, such as the comparative speed of moving vehicles, the number of vehicles, and the broadcasting range of vehicles contribute to communication capability in vehicular CPS. Experimental results show that, when the relative speed of vehicles was higher, then the data size exchanged, and the transmission range was lower.

- Disaster recovery: Natural disasters, such as earthquakes, floods, tsunamis, landslides, etc. can be warned earlier with the help of a CPS-based disaster management system. [16] proposed an automatic control system to interact with physical systems to enable disaster recovery with modules for state recognition, identification of appropriate action and an interface between the software part and the physical part of the CPS.
- Industrial automation: The CPS concept is widely applied to develop fully automated smart factories. [17] designed an efficient industrial system that makes use of CPS principles to handle the widely varying requirements of the various production industries.
- Healthcare: CPS has created a revolution in the healthcare industry. Based on application, CPS nodes can be used to collect clinical trial data, medical expense data, medical images and emotion data (how the patient feels). An adapter is used as a middleware by a CPS node to interact with the system. Adapters also perform the required pre-processing of the data gathered. Distributed File Storage (DFS) techniques provide efficient data storage in health CPS [18] Cloud and big data analytics help in enhancing the performance of smart healthcare systems.

Other applications include smart grid, smart home, smart buildings, smart cities, agriculture, aerospace, water and mine monitoring.

3.4.2 Challenges

- Communication protocols that enhance Quality of Service (QoS): For building a CPS for a real-time application, the protocols required for communication/transmission should be developed such that they are able to provide the desired quality of service for that specific application.
- Resource management: Auto-management techniques need to be developed to resolve resource management issues arising as a result of large volumes of data generated from sensors.
- QoS-aware power management: Central Processing Unit (CPU) energy consumption will be more in CPS due to real-time control.

Hence, there is a need for QoS-aware power management techniques that minimize power consumption.

- Real-Time data management: CPS deals with large amounts of data. Providing real-time data services for CPS in a timely manner is a challenge.

3.5 Design Drivers and Technical Challenges

3.5.1 Coverage and Deployment

The ability to collect the required data from the area being monitored is defined as the coverage ability. Guaranteeing that the area under control is efficiently monitored and that the collected data is properly communicated are the major objectives of coverage in CPS. Avoiding disconnections is a primary parameter as disconnection disturbs the overall performance of CPS, in that CPS requires the network to be connected most of the time. In order to build reliable and predictable CPS, the decision-making systems should successfully receive CPS inputs without interruption.

Deployment means finding the best location to keep the sensor nodes in the network, so that the requirements are well satisfied. In order to ensure that the total coverage is provided by the CPS, the CPS has to check whether the required number of sensors are deployed in the area to be sensed. This becomes the main objective of CPS [19].

The sensor deployment strategies to ensure a wider coverage [20] of CPS are:

- Fixed sensor deployment

  This strategy can be further subdivided into a manual deployment approach and a random deployment approach. For applications such as healthcare, the manual approach of placing the sensor nodes is preferred. Since the number of nodes to be deployed is less and the area of coverage is also less, this approach can perform better than a random deployment scheme. For large-scale networks, random deployment would be the correct choice.

- Deployment of movable sensors

  Movable sensors are used in locations where the field to be observed is very hazardous to reach for a human being. The mobility of the sensor nodes will be used to reach the area of the monitoring field, where no sensor nodes are available to monitor the events occurring in certain locations of the area to be covered.

- Mobile robot deployment

The scheme is well suited for node deployment in a large-scale network as, at the same time, it avoids the node redundancy problem. This method solves the problem of unpredicted obstacles in the pathway of mobile sensors and also checks for coverage holes in the monitoring area.

3.5.2 Mobility

Security, energy consumption, stability and dynamic characteristics of the environment are the important challenges faced by mobile CPS [21]. The difficulties include dynamic network topologies of vehicular mobile CPS, uncertainty in connectivity and resource heterogeneity in the routing protocols. Self-adaptive systems, utility-based forwarding and replication mechanisms are among the approaches to solving mobility challenges. Mobile education and healthcare systems are other examples of mobile CPS.

3.5.3 Energy Efficiency

Energy efficiency is an important factor in a manufacturing environment since it affects the battery life. In order to reduce the energy consumption by physical elements, a green system was developed using advanced software methodologies [22]. A cyberphysical production network, that supports energy-aware manufacturing, was framed [23]. The focus of the framework was to integrate an energy-efficient manufacturing model in a cyberphysical production network.

3.5.4 Data Aggregation

Due to the widespread usage of cyber physical devices, huge volumes of sensor-collected data are received as input by aggregation units. Data are received by aggregators from different types of data sources. Hence, it is very important to check whether the received data are from an authentic source or not, so there is a necessity to develop algorithms to efficiently aggregate sensed data and develop algorithms to identify the location of the monitoring field in a less-secured aggregator. At the same time, the confidentiality of user information also needs to be preserved. The accuracy of an aggregator may be affected when the reply from users is repeatedly sent to the aggregator. Hence, a bit-choosing algorithm was proposed [24] to ensure that the aggregators do not receive repeated user replies, with the help of min, k-th min and percentile computation.

3.5.5 Quality of Service (QoS)

The complex nature of CPS, with different applications operating at different times and on different scales, increase the difficulty in providing end-to-end

QoS for CPS. One of the important aspects in developing a CPS to continuously observe patient health is to provide the service quickly, without any time delay. Network congestion, random loss in network packet and variation in network latency may cause unpredicted behaviour of CPS in emergency service applications. The QoS offered by certain CPS layers of automation systems depends on the quality of data provided by the deployed sensors and the time at which the sensed data is received [25]. The QoS metrics include the sampling rate at which an event is monitored, the network speed by which the received data is communicated to the cloud, data accuracy and quality, the range up to which the environment can be observed, the mobility of the sensing device and the cost.

3.5.6 Resource Management

Efficient resource allocation becomes critical in CPS because of its limited battery energy and memory. Different types of tasks arrive at automation systems that vary with respect to sensing, measurement, computation and processing. The type of task arriving at CPS may be random and CPS may favour certain tasks as a result of its own characteristics. Due to this behaviour, there may be insufficient resources to perform certain jobs which have arrived at random. A CPS that shares the available resources has been developed [26], with an objective to maximize resource utility through a decentralized control.

3.6 Conclusion

CPS applications help tremendously to improve the comfort of our daily life. The technical challenges and research issues in CPS have been addressed in this chapter. The issues discussed show the relative lack of research in the area of security protocol development to withstand security threats, the necessity to combine Wireless Sensor Networks with cyber physical systems and the requirement to improve Quality of Service in various fields of CPS applications.

References

1. Criado, J., Asensio, J., Padilla, N. and Iribarne, L., 2018. Integrating cyber-physical systems in a component-based approach for smart homes. *Sensors, 18*(7), p.2156.

2. ISA95, [Online] Available: https://www.isa.org/isa95/.
3. Lee, J., Bagheri, B. and Kao, H.A., 2015. A cyber-physical systems architecture for industry 4.0-based manufacturing systems. *Manufacturing Letters, 3*, pp.18–23.
4. Ahmed, S.H., Kim, G. and Kim, D., 2013, November. Cyber physical system: Architecture, applications and research challenges. In: *Wireless Days (WD), 2013 IFIP* (pp. 1–5). IEEE.
5. Jiang, J.R., 2018. An improved cyber-physical systems architecture for Industry 4.0 smart factories. *Advances in Mechanical Engineering, 10*(6), p.1687814018784192.
6. Misra, S., Krishna, P.V., Saritha, V., Agarwal, H., Shu, L. and Obaidat, M.S., 2015. Efficient medium access control for cyber-physical systems with heterogeneous networks. *IEEE Systems Journal, 9*(1), pp.22–30.
7. Kang, K.D. and Son, S.H., 2008, June. Real-time data services for cyber physical systems. In: *Distributed Computing Systems Workshops, 2008. ICDCS'08. 28th International Conference on* (pp. 483–488). IEEE.
8. Shakshuki, E.M., Malik, H. and Sheltami, T., 2014. WSN in cyber physical systems: Enhanced energy management routing approach using software agents. *Future Generation Computer Systems, 31*, pp.93–104.
9. Li, H., Qiu, R.C. and Wu, Z., 2012, June. Routing in cyber physical systems with application for voltage control in microgrids: A hybrid system approach. In: *Distributed Computing Systems Workshops (ICDCSW), 2012 32nd International Conference on* (pp. 254–259). IEEE.
10. Liu, T., Yang, L., Lin, Q., Guo, Y. and Liu, Y., 2014, April. Anchor-free backscatter positioning for RFID tags with high accuracy. In: *Infocom, 2014 Proceedings IEEE* (pp. 379–387). IEEE.
11. Jiang, C., He, Y., Zheng, X. and Liu, Y., 2018, April. Orientation-aware RFID tracking with centimeter-level accuracy. In: *Proceedings of the 17th ACM/IEEE International Conference on Information Processing in Sensor Networks* (pp. 290–301). IEEE Press.
12. Kaiwartya, O., Cao, Y., Lloret, J., Kumar, S., Aslam, N., Kharel, R., Abdullah, A.H. and Shah, R.R., 2018. Geometry-based localization for GPS outage in vehicular cyber physical systems. *IEEE Transactions on Vehicular Technology, 67*, p. 5.
13. Shrivastava, A., Derler, P., Baboud, Y.S.L., Stanton, K., Khayatian, M., Andrade, H.A., Weiss, M., Eidson, J. and Chandhoke, S., 2016, October. Time in cyber-physical systems. In: *Proceedings of the Eleventh IEEE/ACM/IFIP International Conference on Hardware/Software Codesign and System Synthesis* (p. 4). ACM.
14. Lisova, E., Uhlemann, E., Åkerberg, J. and Björkman, M., 2017, October. Monitoring of clock synchronization in cyber-physical systems: A sensitivity analysis. In: *Internet of Things, Embedded Systems and Communications (IINTEC), 2017 International Conference on* (pp. 134–139). IEEE.
15. Rawat, D.B., Bajracharya, C. and Yan, G., 2015, April. Towards intelligent transportation cyber-physical systems: Real-time computing and communications perspectives. In: *Southeastcon* (pp. 1–6). IEEE.
16. Mariappan, R., Reddy, P.N. and Wu, C., 2015, December. Cyber physical system using intelligent wireless sensor actuator networks for disaster recovery. In: *Computational Intelligence and Communication Networks (CICN), 2015 International Conference on* (pp. 95–99). IEEE.
17. Kaihara, T. and Yao, Y., 2012, December. A new approach on CPS-based scheduling and WIP control in process industries. In: *Proceedings of the Winter Simulation Conference* (p. 182). Winter Simulation Conference.

18. Zhang, Y., Qiu, M., Tsai, C.W., Hassan, M.M. and Alamri, A., 2017. Health-CPS: Healthcare cyber-physical system assisted by cloud and big data. *IEEE Systems Journal, 11*(1), pp.88–95.
19. Ali, S., Qaisar, S., Saeed, H., Khan, M., Naeem, M. and Anpalagan, A., 2015. Network challenges for cyber physical systems with tiny wireless devices: A case study on reliable pipeline condition monitoring. *Sensors, 15*(4), pp.7172–7205.
20. Lin, C.Y., Zeadally, S., Chen, T.S. and Chang, C.Y., 2012. Enabling cyber physical systems with wireless sensor networking technologies. *International Journal of Distributed Sensor Networks, 8*(5), p.489794.
21. Guo, Y., Hu, X., Hu, B., Cheng, J., Zhou, M. and Kwok, R.Y., 2017. Mobile cyber physical systems: Current challenges and future networking applications. *IEEE Access, 6*, pp.12360–12368.
22. Horcas, J.M., Pinto, M. and Fuentes, L., 2019. Context-aware energy-efficient applications for cyber-physical systems. *Ad Hoc Networks, 82*, pp.15–30.
23. Lu, Y., Peng, T. and Xu, X., 11–13 October 2017. Cyber-physical production network for Energy-Efficient Manufacturing: A framework. In: *CIE47 Proceedings.*
24. Yu, J., Wang, K., Zeng, D., Zhu, C. and Guo, S., 2018. Privacy-preserving data aggregation computing in cyber-physical social systems. *ACM Transactions on Cyber-Physical Systems, 3*(1), p.8.
25. Shah, T., Yavari, A., Mitra, K., Saguna, S., Jayaraman, P.P., Rabhi, F. and Ranjan, R., 2016. Remote health care cyber-physical system: Quality of service (QoS) challenges and opportunities. *IET Cyber-Physical Systems: Theory & Applications, 1*(1), pp.40–48.
26. Nayak, A., Reyes Levalle, R., Lee, S. and Nof, S.Y., 2016. Resource sharing in cyber-physical systems: Modelling framework and case studies. *International Journal of Production Research, 54*(23), pp.6969–6983.

4

Intelligent Building and Environmental Controls for Futuristic Smart Cities Powered by Cyber Physical Intelligence

Rajesh Harinarayan Rajasekaran, Rajkumar Krishnan and Mercy Shalinie Selvaraj

CONTENTS

Organization of the Chapter

Section 1 presents the terms and terminologies to help the reader to understand the chapter. **Section 2** introduces the concept of cyber physical intelligence in building smart cities. **Section 3** discusses the various needs and developments in water management systems for intelligent buildings, whereas **Section 4** reviews various advances in ambience management systems, using cyber physical systems (CPS). **Section 5** describes research into intelligent waste management systems and **Section 6** describes energy management systems in intelligent connected buildings. **Section 7** describes the current research into safety and maintenance of cyber physical intelligent buildings, whereas **Section 8** discusses the various challenges and opportunities in creating intelligent buildings of the future. Finally, the chapter is concluded by **Section 9**.

4.1 Terms and Terminologies

CPS (Cyber Physical Systems): the physical things which are connected through communication media, like the Internet, thus creating CPS.

HVAC (Heating Ventilation and Air Conditioning): the technology of room and vehicular environmental comfort.

IoE (Internet of Everything): bringing people, processes, data and things together to make networked connections to achieve new capabilities, better experience and increased commercial opportunities.

AI (Artificial Intelligence): a branch of computer science that highlights the production of intelligent machines, which think and function like humans.

WSN (Wireless Sensor Networks): wireless networks composed of spatially dispersed independent devices, using sensors to observe the physical and environmental conditions.

Gamification Framework: The application of Game play elements like points, score boards, badges in other areas of activity to encourage participation.

4.2 Introduction

Human shelter has evolved with the progression of various new building materials, from caves, mud huts, to wooden or stone houses, etc. This evolution of material usage in the building of houses occurred on basis of needs, but later building architectures incorporated new facilities, with the use of electronic devices for sophisticated innovations like ventilation, water, energy, drainage systems, etc.

Today's world has the most-sophisticated buildings, as the physical elements have been connected through communication media, such as the Internet, thus creating Cyber Physical Systems (CPS). The buildings of today have many CPS aspects, like smart air cooling systems that are aware of environmental conditions, like room temperature, and take action accordingly, smart water meter systems that measure the usage of water in buildings and provide details to each home and corresponding government bodies for analysis, and smart electric power maintenance systems that measure electricity usage and also automate control to user commands from smart phones.

All these CPS in buildings make our lives more sophisticated but, in order to make it smart, we need to add an intelligence factor. We can achieve cyber physical intelligence when all these smart utilities are connected through an intelligent agent like today's Google Home or Amazon Alexa. The advances in machine-learning domains, like distributed machine learning, Stream Analytics and hardware advances, such as Edge tensor processing units (TPUs), neural sticks and nano graphics processing units (GPUs), can provide the base for connected intelligent agents in buildings. As we connect various multi-modal CPS in smart buildings, that use vision, language and control, to a general intelligent agent, we will move one step towards achieving "Singularity".

The basic needs for human beings are food, water and shelter. Buildings were initially built for shelter and privacy, and they evolved to achieve other purposes. The main reason for buildings is for shelter, that is to protect ourselves from external climatic conditions and physical threats. Once our physical threats from animals were dealt with, our goal for buildings moved towards "Sophistication/Comfort". These more sophisticated buildings were then created through innovative architectural designs and the inclusion of basic electronic devices for heating and ventilation, air coolers, plumbing, etc. [1].

Each industrial revolution brought more sophistication to human lives and increased productivity, creating urban cities. The first industrial revolution started in 1784 with the invention of steam power and the mechanization of new machines. The second industrial revolution began in the 1870s, with the invention of electrical energy and its application to mass production, assembly lines in industry, etc. Both these revolutions have acted as a great stimulus for industry, but they also had a dark side. Rapid urbanization created greater population densities, while the lack of essential resources, like clean water and healthy living places, created health problems in urban migrated people [2]. The third industrial revolution began in the 1950s, with the invention of personal computers, automation and electronics. In this era, consumer electronics boomed and they benefitted the relevant industries, as well as the citizens. As mentioned by C N Trueman [3], inventions like computers, the Internet, sensor systems in aircraft, health and safety, medical electronics, satellite communication systems, washing machines, robot automation, nuclear energy systems, renewable energy systems like solar cells, etc., brought the human way of living to the next level. In this era, our buildings, with basic architectural designs, changed into more sophisticated living spaces. Our buildings were filled with primitive sensor systems, embedded computers, air-cooling machines, machines for human tasks like washing, cleaning, etc., thus providing the highest available level of sophistication. But, even after all these inventions, we still struggle to achieve perfection. So, the next (fourth) industrial revolution, Industry 4.0, is moving towards "Intelligence", that will achieve high levels of adaptation to the ever-changing natural environment we live in. The fourth industrial revolution has started with innovations like cyber physical systems, the Internet of Things, machine learning, etc.

After each industrial revolution, our human lives have also evolved. Our buildings have also evolved from basic to sophisticated buildings to intelligent buildings. An intelligent building's main goals are to monitor, adapt, communicate, control, and optimize living spaces in response to human and environment factors. Our living environment is sophisticated with electronics and sensor machines, yet it has a gap that needs to be filled with human intelligence. Hence, the concept of intelligent buildings was put forward in the 1980s.

There are many vague definitions of intelligent buildings, because we have many types of buildings that constitute an urban city. A sample of the types of buildings in a city is depicted in Figure 4.1.

Each building might have specific intelligent control requirements in order for its intended operations to work smoothly; for instance, a residential building's requirement for 'heating, ventilation and air conditioning' (HVAC) is to maintain room temperature at an ideal temperature, based on the user's comfort/mood, whereas a food storage building in the dairy industry requires the temperature to be less than 2°C. A nuclear/chemical reactor building requires the utmost perfect control of its environment. whereas residential buildings do not require too many sensors nor too much advanced software. Thus, the requirement varies for each building and generalizing a specific set of modules is not possible, so we have generalized definitions. Our generalized definition is that an intelligent building should monitor,

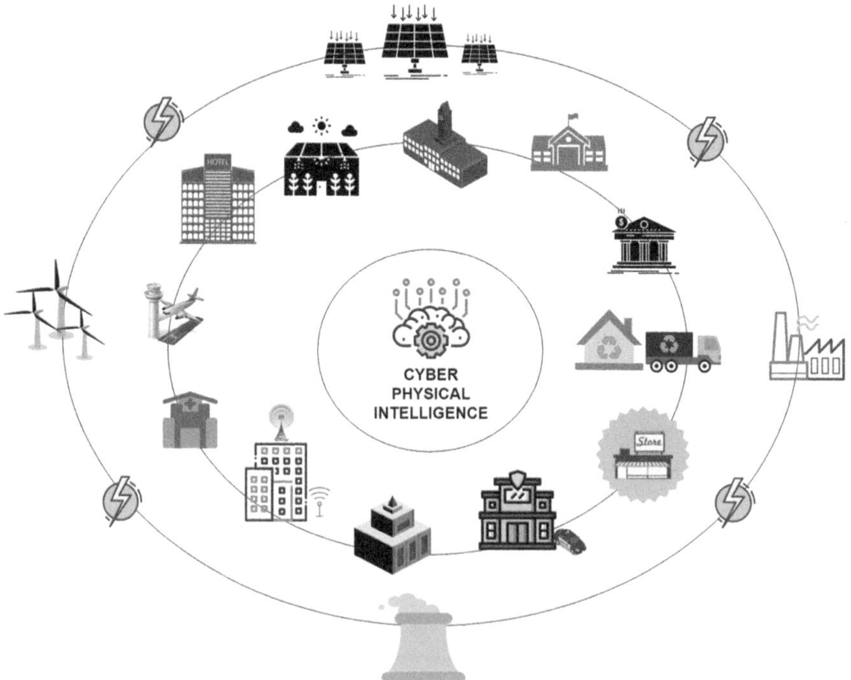

FIGURE 4.1
Intelligent buildings of a smart city powered by cyber physical intelligence

adapt, communicate, control, optimize living spaces in response to human and environmental factors. But the review by Osama Omar [4] devised a list of modules that could be generalized as building blocks of an intelligent building, namely environmental friendliness, space utilization or flexibility, cost effectiveness, human comfort, work efficiency, safety and security, culture, image of high technology, construction process and structure, health and sanitation.

The need for intelligent buildings is essential in the critical environmental situation in which we find ourselves today. Climate change is happening faster, and the Earth's temperature is expected to rise by 1.5°C from 1990 to 2030 [5], and there have been many natural disasters, such as flooding and drought, as well as elevated temperatures. The environment is moving out of our control, and, if we don't manage global warming, then life as we know it will cease to exist. To control these environmental issues, we need to take advantage of the Industry 4.0 concepts, such as CPS and artificial intelligence. This field of application is called environmental controls and Figure 4.2 depicts the environment control features for an intelligent building.

A cyber physical system is a system that contains interacting analogue, digital, human and real-world physical components which are designed for a function, and which work with the help of logical and physical models [6]. The development of CPS has led the industries to the fourth industrial revolution, that is Industry 4.0. They are also the foundational block of the smart city infrastructure. The physical things are connected to the cyber space for automated control and could be effectively used in the intelligent buildings concept, but only when intelligence is added to the CPS in buildings can we arrive at an intelligent building. Machine learning has achieved what

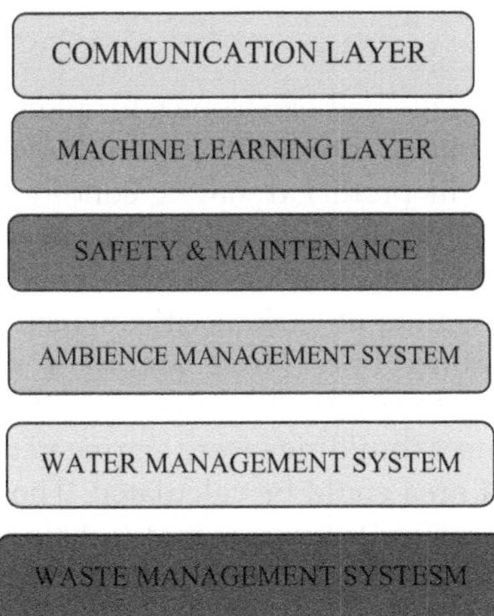

FIGURE 4.2
Environmental control system framework for an intelligent building

even humans could find difficult. When we combine CPS with the Internet of Things (IoT) and artificial intelligence (AI) technologies, we fill the human gap with an artificial intelligent agent that can take intelligent decisions to achieve an optimized living/working space.

4.3 Water Management Systems

Water is one of the most valuable resources for life, and the planet's water resources are in critical condition because they have been used lavishly without proper management. The human population is increasing steadily and the resources are becoming scarce, with Water Day Zero being announced in South Africa, California, Chennai, etc. This extreme situation has been reached because of extreme climatic conditions like greatly reduced rainfall, excessive temperatures, etc., and improper management of natural resources. In the management of water resources in intelligent buildings, we need to check the following details:

- We need to record the quantity we consume to use water resources with caution.
- We need to replenish the water resources by rainwater harvesting and recycling greywater.
- We need to monitor the quality of the water resources supplied to the buildings.
- We need communication alerts in case of fault occurrence in the water system.

Our sophisticated buildings had electronics for plumbing systems or water measurement devices, but an intelligent (smart) building, made with cyber-physical intelligence, will predict dynamic conditions and will adapt. To know about these dynamic conditions, first we need monitoring of water quantity in the buildings and its usage behaviour. Water meter sensor systems have been implemented in various cities which show, in a display, the quantity of water consumed in real time, and these values help the citizens in residential buildings to use water carefully. These water consumption data for individual houses/buildings were aggregated for a city area, so that water demand for that area could be calculated. These water demand data could be used for forecasting scenarios and to help the government water management board with advanced predictions, using machine learning algorithm regression models, deep-learning frameworks, etc., enabling the prevention of drought situations by predicting water conditions and redirecting water from different places.

To manage these water resource levels or water consumption levels, the CPS has been extended to cyber physical social systems. In a cyber physical social system, the physical things of the real world are connected to the cyber space, and the system interacts with humans through behavioural change models. These behaviour change systems have been used in water management systems by providing personal usage level statistics/dashboards, and a gamification framework on top of it.

In Aslansefat et al. (2018), resilient systems, created using the Markov modelling technique, were used in the behaviour change system to encourage guests in a hotel to be more water cautious [7]. The aim of this research is to reduce water consumption by creating an awareness in the guests and also by the process of anomaly detection. The case study is a hotel and three levels of water consumption are defined: low, moderate and high usage. The usage data is sent to the cloud and aggregated. The paper also differentiates between an alert alarm resulting from machine failure or human-made failure, by using the estimated human behaviour. So, if there is a water leakage in a pipe, the system alerts the hotel managers and distinguishes between whether it's a machine failure or human-made failure [7].

Further research, by Curry et al. (2018), shows how smart water monitoring could be carried out in various facilities, such as smart homes, schools, office buildings and airports [8]. The aim of the research was to create awareness to various users about the eco-friendly usage of water and energy, with the help of a cyber physical social system framework. Through gamification of these tasks, like providing leader boards, rewards, badges, etc. the citizens were influenced to show eco-friendly behaviour [8].

Stoyanav et al. (2018) [9] introduced a new concept for monitoring water resources through cyber physical social systems. An intelligent building could be placed in various land types like urban, mountain, seaside, etc., and is expected to provide intelligence services without any difficulty. In this paper [9], there is a need to manage water resources in a seaside city in Bulgaria. But seawater, as a "thing", is difficult to model with a cyber physical system, because the sea occupies such an enormous spatial zone. To model it, the paper uses Virtual Physical Space (ViPS) architecture, that contains ontology, events, temporal and ambient factors. This research method, using ViPS, provides dynamism, flexibility and intelligence.

These intelligent buildings do not always have the luxury of tapping into naturally available water resources like groundwater or storm water. Most buildings in a city are supplied with water by the municipalities, which plan water distribution based on area-wise water demand. These interconnected water resources, through cyber physical intelligence, are called "smart water grids". Water distribution networks are very large, and deploying sensor nodes is difficult and costly. As a consequence, simulation tools have been created to predict areas that might be susceptible to damage, like the EPANET [10], or to use improved wireless sensor network (WSN) architecture, such as PIPENET [11], to detect faults or water leaks in the distribution network lines. These WSN architectures are programmed to achieve goals like maximization of accuracy, minimization of detection time, or expansion of the coverage area. With such sensing systems, connected to Information Communication Technologies (ICT) in place, the quantity, quality and faults could be monitored in the connected water distribution network for various buildings within a smart city. The existing smart water grids in today's world [12] are:

- Southeast Queensland (SEQ) water grid in Australia.
- Water supply network department (PUB) in Singapore.
- National Smart Water Grid (NSWG) in the USA.
- Smart Water Grid (SWG) research in Korea.

Without adequate supplies of water, human life will perish, but it will be in greater danger if low-quality water is provided to them. Water quality monitoring is very important in intelligent buildings, where any form of water, such as rainwater, groundwater, recycled water, or municipally supplied water, must be quality tested. Cyber physical quality monitoring systems, with quality measurement sensors, could be used to provide real-time quality monitoring and communication. Water quality could be jeopardized by unwanted chemical elements, like lead or arsenic, or biological elements, such as viruses and bacteria.

Water quality should be tested continuously at the source of water, like rivers, ground water or building storage facility. It should also be tested when distributing the water in pipelines to various buildings in the city. Cyber physical water quality monitoring systems would provide better insights than traditional quality measurement methods, as mentioned in research by Bhardwaj et al. (2018) [13]. The paper proposes a cyber physical water quality system, using the fuzzy logic technique in a water distribution network.

4.4 Ambience Management Systems

Air is also a natural resource essential for life to thrive, but, unlike water, air is abundant in our world. After the formation of our atmosphere, life began to replicate under the now-favourable conditions. Living organisms respire by inhaling oxygen from the air and releasing unwanted gases, like CO_2. The problem here is 'Climate Change'; the Earth has seen extreme climatic fluctuations throughout history from its creation, and through the Ice Ages, until the global warming phase which we are currently enduring. Life has always found a way to adapt to climate changes, either naturally, like a camel that can regulate its body temperature during the day and night by adapting to the desert environment, or by man-made designs, such as the igloo, created by Inuit, which provided warm shelters in ice-cold polar regions. Then, with each industrialization phase, we added more pollutants to our atmosphere and aggravating living conditions; polluted air is bad for living organisms and greenhouse gases increase the Earth's temperature, as well as affecting water resources. So, for a system to control such variable climatic conditions, we need an intelligent shelter/building system that can:

- Measure, analyze and report the air quality in the environment under study
- Measure, analyze, predict and report air temperature in the environment under study

To achieve the above requirements, we need CPS designs, because they can continuously monitor the environment with sensors, analyze and pre-process the data, and report the data in a user-readable form through dashboards, visualizations, etc. There have been various implementations of indoor air quality monitoring in buildings, such as iAQ, iAQ mobile, iAQIoT, iAirC, and iDust systems, each measuring various air quality parameters, such as temperature, humidity, CO_2, CO and particulate matter. These systems not only sense air quality, they are also designed with a network data pipeline to send air quality data to servers for analysis and also to end-user devices, like mobile apps or websites. In this design, the building manager is the end user, and this set of insights from the air quality CPS would assist the manager in policy changes and behaviour management [14].

Since intelligent buildings will be fitted with more sensors, the amount of data streams will be huge, but this could be a great advantage for cyber physical intelligence systems that require more data to produce valuable insights. Stamatescu et al. (2019) [15] proposed a data mining pipeline that could take in huge data streams and provide results for decision making and control for smart ventilation systems in a building [15]. A two-step data processing framework is used, the first phase compressing the huge raw data streams by using the Aggregate Approximation method, with the resulting streams being inputted to a Support Vector Machine (SVM) algorithm that classifies the data stream to appropriate ventilation units [15].

Artificial intelligence applications have reached everywhere, due to the development of 'Deep Learning' algorithms. Du et al. [16] proposed a new Deep Learning Air Quality Forecasting Framework (DAQFF), which outperforms other air quality learning frameworks, such as Convolutional Neural Network (CNN), Recurrent Neural Network (RNN), Gated Recurrent Unit (GRU), Long Short-Term Memory (LSTM) and Support Vector Regression (SVR). The training of the proposed model was carried out on the Beijing

PM2.5 dataset and the urban air quality dataset. The DAQFF framework has two major learning components, the one-dimensional convolutional neural network (CNN) and the bi-directional long short-term memory (LSTM). Since multivariate time series data are provided as input, fusion learning is also presented as a layer.

Increasing the numbers of sensors in the building will improve accurate measurement, but will adversely affect the energy requirement of the building. To reduce the amount of energy consumed by the deployment of multiple sensor modules, the concept of human-in-the-loop systems to crowd source data is proposed. The crowd-sourced measurement of indoor radon gas concentration in granitic buildings is a target of such research, as radon gas is one of the main causes of lung cancer and is present in granitic buildings. The proposed human-in-the-loop CPS air quality system consists of a cloud app engine to store and carry out cloud-based reasoning on the data collected from the end devices. The end sensor devices are geo-referenced to obtain spatial details, along with air quality data like radon concentration, temperature, humidity and CO_2 levels. The connection between all the sensor nodes and connectivity to the cloud engine is provided using a Long Range Wide Area Network (LoRaWAN) architecture. The insight result is sent to the human users to manually close or open ventilation units, thus allowing the human-in-the-loop CPS to mitigate air quality problems. In addition, the results are connected for automatic control of ventilation units if the air quality rules provided by the manager are matched [17], creating a closed loop human-in-the-loop CPS for indoor air quality, maintaining low energy usage.

The above-mentioned research was carried out in public buildings made of granite construction; there has also been research that good indoor air quality will actually boost the performance of the workers as well as their health [18]. So, improving air quality in intelligent buildings does not just help to save the Earth and make humans comfortable, it also improves the productivity of humans and adds economic benefits to organizations.

4.5 Waste Management Systems

Waste management is an essential environmental control system for smart cities, because unprocessed waste dumping can lead to climate change by releasing harmful gases and, in turn, accelerating global warming, thereby further evaporating water resources. Thus, this chain-of-life connected resource is hanging in the balance; when compromised, this creates a chain-of-disasters. The amount of wastage is increasing dramatically; for instance, the landfills in Delhi are predicted to grow to more than the height of the Taj Mahal [19].

Each building type, be it a school, college, residence, hospital or factory, produces waste in large quantities, as the population within each building is large. The wastes are of different types, such as liquid, solid, organic, recyclable and hazardous wastes [20]. Each type of waste requires different recycling or reuse strategies; government municipalities are in charge of waste treatment in a city, but waste could be reduced if citizens or building occupants are aware of the problem.

Food wastage is high in today's world, as the number of mouths to feed has increased. Food is wasted at all stages of the food chain, from agricultural production, in transit, in markets and in residential buildings or hotels, where it is ultimately consumed. Food waste comes under the category of organic waste, and it could be recycled into compost or manure (after feeding to farm animals) for farms, thus avoiding it ending up in landfills. But we all know that prevention is better than cure, and we could reduce food waste if we could predict the demand for food. This prediction task is hard for a human, but a cyber physical intelligent system can predict, using data and machine learning algorithms. The research by Sakoda et al. (2019) [21] provides optimized stock management through demand estimation. The demand estimation is achieved through statistical models that are inputted with point-of-sales data. Through this research, food waste is avoided to a large extent, while the economic benefit to the target building is also increased.

The above-mentioned research uses only data science methods to estimate the demand for retail products, but, to obtain real-time data and make the entire process cyberphysically intelligent, we need sensor systems in retail stores. In today's world, this concept is implemented by the company Amazon, which has ventured into the retail markets with its AmazonGo system [22]. The system has an open-door policy system, that monitors each user continuously in terms of what products the user takes and calculates the total cost in real time. The users do not have to wait in a queue to pay their bills; they can directly walk out of the store and the amount will be automatically deducted from their connected bank account. Thus, it collects real-time data, avoids congestion within the store, and provides economic benefit for the store owner. This AmazonGo model is a fitting example for cyber physical intelligence.

Liquid waste, like human faeces or sewage, could not be treated by each building locally, as it would be prohibitively expensive to build treatment plants in each building. But each building could have greywater recycling designs; greywater is the water that could be reused, with minor treatment, such as rainwater, dishwasher water, shower water, etc. But the problem and exploitation of cyber physical intelligence in this context is similar to the water management tasks, like quality measurement, leak detection, etc., mentioned in Section 4.3.

Waste is inevitable; even though some waste could be recycled, the remaining waste should be treated at different government facilities. But the problem is that the citizens or occupants of the building might not eco-friendly, failing to collect and/or segregate the waste from their buildings. This waste has to go from bins in the buildings to the treatment site through logistics; this whole chain could be consecutively managed using various IoT waste management designs for smart cities. The IoT design, described by Abdullah et al. (2019) [23], provides an overall monitoring of the waste chain, using smart bins, with GPS-deployed logistics for planning the collection of waste in a timely manner.

But this research does not make citizens feel responsible for recycling and change their behaviour, so, to encourage such positive behaviour, various gamified waste management systems have been developed. Briones et al. (2018) [24] provides a gamification framework for encouraging the citizens to take part in the waste recycling process, whereby the citizens are rewarded for their deposit of waste into the smart bins. Then, the usual framework of connected bins and waste collection trucks is automated. The volumetric sensors placed in the smart bin measures the level status of the bin and, through low-band IoT communication, details of the bins ready for emptying are communicated for collection. Weight sensors are used in smart bins to measure the weight of the waste the user has provided and to reward them accordingly. Thus, these gamification cyber physical methods promote eco-friendliness and improve the efficiency of waste management.

But waste collection is not the only step in waste management; there are other important steps, like waste segregation, waste recycling, etc. Tasks, such as waste segregation, are still made by humans and are prone to error, so we can take advantage of the cyber physical intelligent waste classification as proposed by Bobulski and Kubanek (2019) [25]. The proposed waste classification framework uses Convolutional Neural Networks (CNN) to classify plastic wastes into polyethylene terephthalate, polypropylene, high-density polyethylene and polystyrene. The customized AlexNet version classification algorithm performs better in this multi-class classification problem, with a high accuracy greater than 95%.

The above research provides only the intelligence, but, if it is connected to a CPS, the waste segregation task could be automated. Research by Sreelakshmi et al. (2019) [26] has produced a cyber physical intelligent waste-segregating machine that classifies plastic and non-plastic waste. The classification algorithm used by the machine is Capsule Neural Networks (CapsuleNet). The CapsuleNet was trained on a privately collected dataset and it performed with a classification accuracy of 96.3%. The classification model was written in Tensorflow with a Keras background and trained on Graphical Processing Unit (GPU). The hardware consists of a webcam and a gate, run by a stepper motor. The stepper motor pushes the waste into plastic or non-plastic bins, based on the classification results.

4.6 Energy Management Systems

Energy management is an important aspect that should be handled throughout the environmental control systems stack. Only a self-sustaining system could provide optimized automation to CPS, because buildings consume more electricity than any other element in a city, more than cars, streetlights, etc. The demand for electric energy is increasing steadily, with building energy consumption in India estimated to grow faster than that in other nations, according to the US Energy Information Administration [27];these data also agree with the 2017 India energy statistics report by the Government of India [28] and the trend is apparent in Figure 4.3.

To make a building self-sustainable, there are many renewable energy sources, and solar power harvesting, using photovoltaic cells on the roof, has started to expand rapidly. The design proposed by Yadagani et al. (2019) [29] combines the solar power generated by photovoltaic cells present on the residential building, along with grid power. The consumption of each device in the house for each user is calculated using a smart meter. Various loads are analyzed and a user-load profile is created, which is used for demand response. If there is an energy requirement more than that produced by the solar cells, then, based on estimated load profile, grid power is requested.

The above is an example of smart metering IoT design, which takes the data and predicts at a different place, like the cloud or a separate server. This might create latency in data transfer, and the real-time analysis for critical

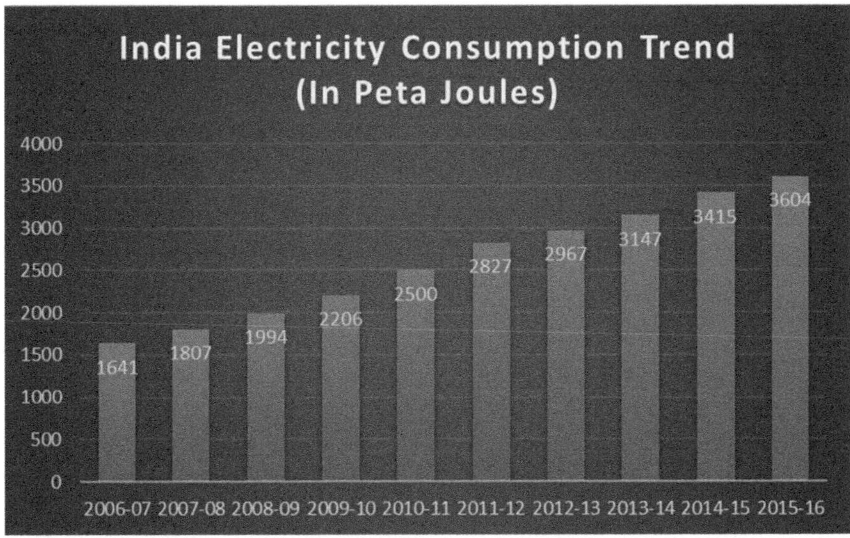

FIGURE 4.3
Electricity consumption according to the India Energy Statistics Report, 2017 [28]

applications might not accept such latency. Research by Buzachis et al. (2019) [30] proposes a microservice-based smart meter, that computes results in the edge devices, instead of sending the data to cloud. The edge computing IoT-as-a-service for smart metering avoids the latency created by transmission to the cloud and provides real-time results. A Fast Fourier transform algorithm is used to characterize the non-linear loads from the grid, and the results are used to protect the appliances in a college campus building, thus reducing faults and ensuring harmonic loads.

Research by Nguyen et al. (2018) [31] provides an overview of the concept of the Internet-of-Everything (IoE) to create a Building Energy Management (BEM) system. This paper shows how everything in the future is going to run on electricity and how we can create an overall connected management system through the Internet-of-Everything. It also describes that buildings consume about one-third of the energy they produce, and proposes that IoT designs will help in building Net-Zero Energy Buildings (NZEBs).

NZEBs should not only maintain power, but also harvest it. Huang et al. (2017) [32] proposes an adaptive scheduling algorithm for energy harvesting sensor networks to make greener cities. The algorithm computes the optimal threshold value for scheduling, based on a reward mechanism. Depending on the reward in the transmitted packet and the current state of the harvested energy, the decision is made in the sensor network power transmission. When we take advantage of harvesting systems in creating net-zero buildings, we will eventually achieve Net-Zero Plus Buildings, which produce more than they consume.

4.7 Safety, Security and Maintenance

Once the indoor environment is controlled through CPS and intelligent buildings are created, we need to protect the infrastructure from internal and external factors. The external factors damaging the infrastructure would be natural factors, such as earthquakes, cyclones, floods, drought, etc., and human factors, like theft, unauthorized entry, etc. Most of the natural disasters can only be predicted but not mitigated with sensor systems. The mitigation can only be done with innovative civil designs.

The prediction can be achieved through cyber physical intelligence systems like predicting earthquakes using the long short-term memory (LSTM) technique, as proposed by Wang et al. (2017) [33], using spatiotemporal data. Flash floods are a common phenomenon nowadays in developing countries and predicting their occurrence could help citizens mitigate the effects. Hiroi and Kawaguchi (2016) [34] proposed a multi-sensor system for monitoring real-time water level and localized heavy rain, predicting flash floods on the basis of regression models.

Landslides area major problem for buildings on a hilly site; research by Liu et al. (2019) [35] was able to detect landslides through sensor data. The proposed CPS design uses Wi-Sun acceleration sensors and communication technology. The design also reduces data traffic by determining correlations between parameters from the collected sensor data, thus saving the energy used and reducing the complexity of the system.

An internal factor in disaster management would be the malfunctioning of systems in the buildings. To resolve this, regular maintenance and timely audit are required. Formation of cracks in buildings is a major problem to be solved, although it is usually not detected at an early stage. As with healthcare, the health of a building should also be checked, and early detection will always avoid the development of serious problems. Gopalakrishnan et al. (2018) [36] proposed a crack detection method, using an Unmanned Aerial Vehicle (UAV). The UAV has an inbuilt camera that can identify cracks on a building with the help of an inbuilt pre-trained deep learning model. Transfer learning was used, along with logistic regression and a single-layer neural network, yielding up to 90% accuracy. Similarly, face-recognition deep learning models have been the most successful application of computer vision, and they are now widely used in surveillance. An intelligent building should deploy such face recognition model surveillance for control of access and thief identification.

But, in the future, when every physical operation is connected in cyberspace, all the systems will be targets for cyber physical attacks. Any manipulated attack on such connected systems could wreak havoc in the world. As we have seen with smart water grids in Section 4.2, an attack on such a connected water infrastructure would be devastating. So, to protect from such attacks, a cyber physical intelligence framework was proposed by Bakolas et al. (2019) [37]. The research describes the use of adaptive deep learning models, with the help of data fusion from heterogeneous sensors, as a solution for protecting water infrastructure to protect against both physical and digital attacks. Intelligent buildings generate a lot of data and so privacy of those data is a major concern. Research by She et al. (2019) [38] provides a possible solution, using blockchain technology. Simulation of blockchain was done in HyperLedger Fabric 1.2.0, and various attack experiments were conducted to validate the strength of the proposed privacy architecture for smart buildings.

4.8 Opportunities and Challenges

The greatest challenge in creating Intelligent Buildings (IB) would be "interoperability" [39]. With so many sensor and hardware manufacturers in today's world, creating a cyber physical intelligence model that could connect all heterogeneous devices would be a major challenge. This offers more opportunities for the currently growing field of containerized machine learning.

Another challenge is securing the privacy of a building user's data. Recently, there have been so many concerns over facial recognition. As a nation, China has been reported to be using AI surveillance to maintain strict security, and this initiative by the Chinese government has provoked mixed emotions. Machine learning privacy and federated learning will also be gaining greater importance in the upcoming AI era.

Handling big data is a major challenge in connected buildings, which provides opportunities for microservices and edge/fog computing research. In order to transfer live-sensed big data, techniques like compressed sensing are used. There are also challenges for creating simulations for connecting buildings in smart cities, because the current simulation tools, like NS2 and OPNET, can simulate only limited numbers of nodes and cannot be used to completely simulate a smart city application [40].

Connected buildings require better network media for transmitting a live stream of sensor data. There have been many problems, such as congestion, latency and availability. The research on 5G software for faster sensing will play a major part in achieving the dream of a smart city.

4.9 Conclusion

This chapter discusses various research into environmental controls, like those of water, ambience, waste, energy, safety management systems, and connecting these control systems for the creation of intelligent buildings. Various applications of artificial intelligence techniques on CPS to create cyberphysical intelligence are also discussed. When the challenges and opportunities mentioned here are solved and exploited, when a general intelligence is created on a smart city buildings scale, when human interaction in the connected world is seamless, then we will have reached closer to "Singularity".

References

1. Oremus, Will. 2013. A history of air conditioning from ancient mountains of snow to the window units of today. https://slate.com/culture/2013/07/a-history-of-air-conditioning.html [last accessed date: 06/07/2019].
2. Industrial_revolution, Jul 2019. https://www.history.com/topics/industrial-revolution/industrialrevolution [last accessed date: 06/07/2019].

3. Trueman, CN. Inventions 1900 to 1990, https://www.historylearningsite.co.uk/inventions-and-discoveries-of-the-twentieth-century/inventions-1900-to-1990/ [last accessed date: 05/07/2019].
4. Omar, Ossama. 2018. Intelligent building, definitions, factors and evaluation criteria of selection. *Alexandria Engineering Journal*, 57(4). doi:10.1016/j.aej.2018.07.004.
5. Earth's temperature to rise 1.5C asearly as 2030 amid dire warnings from UN climate panel, *The Telegraph*October 8, 2018. https://www.telegraph.co.uk/news/2018/10/08/earths-temperature-rise-15c-early-2030-amid-dire-warnings-un/ [last accessed date: 06/07/2019].
6. NIST Cyber physical systems page. NIST, https://www.nist.gov/el/cyber-physical-systems [last accessed date: 06/07/2019].
7. Aslansefat K., Ghodsirad M.H., Barata J., Jassbi J. (2018) Resilience Supported System for Innovative Water Monitoring Technology. In: Camarinha-Matos L., Adu-Kankam K., Julashokri M. (eds) Technological Innovation for Resilient Systems. DoCEIS 2018. IFIP Advances in Information and Communication Technology, vol 521. Springer, Cham. https://doi.org/10.1007/978-3-319-78574-5_7
8. Curry, E. et al. 2018. Internet of things enhanced user experience for smart water and energy management. *IEEE Internet Computing*, 22(1), pp. 18–28.
9. S. Stoyanov, D. Orozova and I. Popchev, "Internet of Things Water Monitoring for a Smart Seaside City," 2018 20th International Symposium on Electrical Apparatus and Technologies (SIELA), Bourgas, 2018, pp. 1–3, doi: 10.1109/SIELA.2018.8447084.
10. Rossman, L.A. 2000. EPANET 2: User's manual.
11. Ivan Stoianov, Lama Nachman, Sam Madden, and Timur Tokmouline. 2007. PIPENETa wireless sensor network for pipeline monitoring. In Proceedings of the 6th international conference on Information processing in sensor networks (IPSN '07). Association for Computing Machinery, New York, NY, USA, 264–273. DOI:https://doi.org/10.1145/1236360.1236396
12. Lee, S.W., Sarp, S., Jeon, D.J. and Kim, J.H. 2015. Smart water grid: The future water management platform. *Desalination and Water Treatment*, 55(2), pp. 339–346.
13. Bhardwaj, Jyotirmoy, Gupta, K. and Gupta, Rajiv. 2017. Towards cyber physical era: Soft computing framework based multi-sensor array for water quality monitoring. *Drinking Water Engineering and Science Discussions*, 11(1), pp. 1–7. doi:10.5194/dwes-11-9-2018.
14. Gonçalo Marques and Rui Pitarma (November 5th 2018). Indoor Air Quality Monitoring for Enhanced Healthy Buildings, Indoor Environmental Quality, Muhammad Abdul Mujeebu, IntechOpen, DOI: 10.5772/intechopen.81478
15. G. Stamatescu, I. Stamatescu, N. Arghira and I. Fagarasan, "Data-driven modelling of smart building ventilation subsystem", Journal of Sensors, vol. 2019, pp. 14, 2019.
16. Du, S., Li, T., Yang, Y. and Horng, S.J. 2018. Deep air quality forecasting using hybrid deep learning framework. *arXiv Preprint ArXiv:1812.04783.*
17. Rosado da Cruz, António Miguel, Lopes, Sérgio, Moreira, Pedro, Abreu, Carlos, Silva, J.P., Lopes, Nuno, Vieira, Jose and Curado, António 2018. On the design of a Human-in-the-Loop Cyber-Physical System for online monitoring and active mitigation of indoor Radon gas concentration. doi:10.1109/ISC2.2018.8656777.

18. Wargocki Pawel, Improving indoor air quality improves the performance of office work and school work and provides economi benefits, 8th International Conference for Enhanced Building Operations - ICEBO'08 Conference Center of the Federal Ministry of Economics and Technology Berlin, October 20 - 22, 2008.
19. Garbage mountain at Delhi's Ghazipur landfill to rise higher than Taj Mahal by 2020. *Hindustan Times*, https://www.hindustantimes.com/india-news/garbage-mountain-at-delhi-s-ghazipur-landfill-to-rise-higher-than-taj-mahal-by-2020/story-RC0kwZdUmdHHfDs3rJGngI.html [last accessed date: 06/07/2019].
20. There are 5 types of waste, do you know them all? 4waste removals pty ltd., December 20, 2016. https://4waste.com.au/rubbish-removal/5-types-wast e-know/ [last accessed date: 06/07/2019].
21. Sakoda, G., Takayasu, H. and Takayasu, M. 2019. Data-science-solutions-for-ret ail-strategy-to-reduce-waste-keeping-high-profit. *Sustainability*, 11(13), p. 3589.
22. Polacco, A. and Backes, K., 2018. The amazon go concept: Implications, applications, and sustainability. Journal of Business and Management, 24(1), pp.79–92.
23. Abdullah, N., Alwesabi, O.A. and Abdullah, R. 2019. IoT-basedsmartwastem anagementsysteminasmartcity. In: *Recent Trends in Data Science and Soft Computing. IRICT 2018. Advances in Intelligent Systems and Computing*, vol. 843, Saeed F., Gazem N., Mohammed F., Busalim A. (eds). Springer, Cham.
24. Briones, A.G. et al. 2018, August. Useofgamificationtechniquestoencouragega rbagerecycling, asmartcityapproach. In: *International Conference on Knowledge Management in Organizations* (pp. 674–685). Springer, Cham.
25. Bobulski, J. and Kubanek, M., 2019, June. Wasteclassificationsystemusingimage processingandconvolutionalneuralnetworks. In: *International Work-Conference on Artificial Neural Networks* (pp. 350–361). Springer, Cham.
26. K. Sreelakshmi, S. Akarsh, R. Vinayakumar and K. P. Soman, "Capsule Neural Networks and Visualization for Segregation of Plastic and Non-Plastic Wastes," 2019 5th International Conference on Advanced Computing & Communication Systems (ICACCS), Coimbatore, India, 2019, pp. 631–636, doi: 10.1109/ICACCS.2019.8728405.
27. Buildings energy consumption in India is expected to increase faster than in other regions, US Energy Information Administration, 2017 October 10, https://www.eia.gov/todayinenergy/detail.php?id=33252 [last accessed date: 06/07/2019].
28. Energy Statistics report, Ministry of Statistics and Programme Implementation government of India, 2017. http://www.mospi.gov.in [last accessed date: 06/07/2019].
29. J. Yadagani, P. Balakrishna and G. Srinivasulu, "An Effective Home Energy Management System Considering Solar PV Generation," 2019 IEEE International Conference on Sustainable Energy Technologies and Systems (ICSETS), Bhubaneswar, India, 2019, pp. 1–6, doi: 10.1109/ICSETS.2019.8744780.
30. A. Buzachis, A. Galletta, A. Celesti, M. Fazio and M. Villari, "Development of a Smart Metering Microservice Based on Fast Fourier Transform (FFT) for Edge/Internet of Things Environments," 2019 IEEE 3rd International Conference on Fog and Edge Computing (ICFEC), Larnaca, Cyprus, 2019, pp. 1–6, doi: 10.1109/CFEC.2019.8733148.

31. V. T. Nguyen, T. Luan Vu, N. T. Le and Y. Min Jang, "An Overview of Internet of Energy (IoE) Based Building Energy Management System," 2018 International Conference on Information and Communication Technology Convergence (ICTC), Jeju, 2018, pp. 852–855, doi: 10.1109/ICTC.2018.8539513.
32. Huang, L., Bi, S., Qian, L.P. and Xia, Z. 2017. Adaptive-scheduling-in-energy-harvesting-sensor-networks-for-green-cities. *IEEE Transactions on Industrial Informatics*, 14(4), pp. 1575–1584.
33. Q. Wang, Y. Guo, L. Yu and P. Li, "Earthquake Prediction Based on Spatio-Temporal Data Mining: An LSTM Network Approach," in IEEE Transactions on Emerging Topics in Computing, vol. 8, no. 1, pp. 148–158, 1 Jan.-March 2020, doi: 10.1109/TETC.2017.2699169.
34. K. Hiroi and N. Kawaguchi, "FloodEye: Real-time flash flood prediction system for urban complex water flow," 2016 IEEE SENSORS journal, Orlando, FL, 2016, pp. 1–3, doi: 10.1109/ICSENS.2016.7808626.
35. Liu, Zhi, Tsuda, T., Watanabe, H., Ryuo, S. and Iwasawa, N. 2019. Data driven cyber-physical system for landslide detection. *Mobile Networks and Applications*, 24(3), pp. 991–1002.
36. Gopalakrishnan, Kasthurirangan et al. 2018. Crack damage detection in unmanned aerial vehicle images of civil infrastructure using pre-trained deep learning model. *International Journal of Traffic and Transport Engineering*, 8(1), pp. 1–14.
37. Bakalos, Nikolaos et al. 2019. Protecting water infrastructure from cyber and physical threats: Using multimodal data fusion and adaptive deep learning to monitor critical systems. *IEEE Signal Processing Magazine*, 36(2), pp. 36–48.
38. She, Wei et al. 2019. Homomorphic Consortium Blockchain for Smart Home system sensitive data privacy preserving. *IEEE Access*, 7, pp. 62058–62070.
39. Lilis, Georgios, Conus, G., Asadi, N. and Kayal, M. 2017. Towards the next generation of intelligent building: An assessment study of current automation and future IoT based systems with a proposal for transitional design. *Sustainable Cities and Society*, 28, pp. 473–481.
40. Yang, Q. Internet of things application in smart grid: A brief overview of challenges, opportunities, and future trends. In Smart Power Distribution Systems; Elsevier: Amsterdam, The Netherlands, 2019; pp. 267–283.

5

Smart Agricultural – An Application of Cyber Physical Systems

G.R. Karpagam, M. Syed Hameed and K. Eshwar

CONTENTS

Organization of the Chapter

Section 1 presents the terms and terminologies to help the user to understand this chapter. **Section 2** introduces the concepts of the cyber physical systems (CPS) and precision agriculture, what CPS means in the present context, and the technologies involved. **Section 3** explains the context of the work involved with respect to smart agriculture, and the approach in which the context can be realized. **Section 4** briefs on the implementation of a prototype based on smart agriculture. It further explains how an input image is recognised, the results are stored and how a solution is provided at the end. **Section 5** extends on how this concept can be explored further and provides an insight into what might be possible in the future.

5.1 Terms and Terminologies

Precision Agriculture – It is a concept of farming that is based on measuring, observing and responding to variability in crops. It uses information technology to ensure that the crops are managed correctly for optimum health and productivity.

Global Positioning System (GPS) – It is a system to determine the position of an object in real time.

Neural Network – It is a series of algorithms that recognise relationships in a set of data, basically mimicking the functionality of the human brain. They can adapt to changing input.

Convolution – The initial step that is involved in identifying the input of a neural network. It involves the input image, the feature that is going to be detected, and the feature map, based on the detected feature.

Optical Character Recognition – It consists of recognising text from an image. The recogniser would be able to obtain the text from the corresponding image.

5.2 Introduction to Cyber Physical Systems (CPS)

CPS can be considered as smart systems that contain the domain of both the physical as well as computational (hardware and software) components. They provide seamless integration and operate in close interaction in order to sense the ever-changing state of the real world. CPS are the result of the integration of software, computer, electronic and motor control – the culmination of all the above fields would result in mechatronics. They can be described as engineering processes, which are transdisciplinary. CPS are equipped with a high level of complexity in terms of various temporal and spatial scales, and also involve highly networked communication.

CPS are those systems that provide the ability to connect the physical world with the virtual world, that is completely involved with the processing of information. There are several contexts in which CPS can exist, one of which is the agricultural industry. When agriculture combines with CPS, this is known as precision (smart) agriculture, which can be defined as a methodology of enhanced decision making, which associates the effective management of farm practices, together with sensing devices, positioning systems from satellites and the use of computers to provide information. The cultivation of crops plays a very important role in agriculture. However, proper maintenance and management is needed. Lack of such management would lead to loss of crops or, at least lowered yields. For example, the ability to detect diseases in an infected crop is a problem of increasing concern. Real-time disease detection systems do not exist in the current scheme of things as they involve large-scale processing time, a tremendous amount of work and plant disease expertise. The proposed precision farming system aims at predicting the diseases present on infected plants in real time. We use the vggnet architecture of ImageNet in order to predict the diseases, using a Convolutional Neural Network (CNN). Disease detection steps include pre-processing of the input images, feature extraction from the infected areas and identification of the plant disease. In addition to this, the proposed solution aims at assisting the farmers by treating with the relevant fungicide from an agrochemical supplier. By this way, a great deal of manual work is considerably reduced.

5.3 Precision Agriculture

Precision agriculture is an effective farm management system which is based on information and technology. It is used in order to efficiently inculcate the very concepts of identification, analysis and management of crop parameters exhibiting field variability, in order to obtain the optimum level of management of resources, sustainability and profitability. In this type of agriculture, the important decision-making situations are left to the information technologies, which help in arriving at an optimal solution. Precision farming targets the increased efficiencies that can be obtained by understanding and dealing with the natural variability found within a crop field. The goal is to distribute inputs and management on a site-specific basis to leverage the yield and to maximize long-term benefit, instead of focusing to apply the same amount of resources everywhere throughout the field. Applying the same techniques and procedures across the field, which was the strategy applied in earlier days, may no longer be the best choice in the present context. Precision farming techniques aim to leverage the effectiveness of crop inputs as much as possible, thereby helping farmers worldwide.

The most essential elements for precision agriculture are data and information. Information can be obtained from technologies such as global positioning systems, global information systems, remote sensing and other technologies. Eventually, those farmers who use information effectively will earn higher returns. In precision farming, customization of smaller areas within fields is possible which makes it distinct from typical agriculture. Precision agriculture is a systems approach to farming and a good farm management system is essential to obtain good results. The problems related to precision agriculture are perceived, rather than realized, benefits and the barriers which minimize the adoption of precision agriculture management on a larger scale. However, the definition of precision farming is suitable only when the land holdings are vast and variability exists between fields. In India, the average land holdings are very small, even with large-scale farmers. In the context of Indian farming, precision agriculture can be defined as precision application of agricultural inputs, based on soil type, climatic and weather conditions and the requirements of the crop, to increase yield and productivity. Today, farmers are looking for new ways to increase efficiency and cut costs because of increasing input costs and decreasing commodity prices. Thus, precision farming technology would be a suitable and adaptable alternative to achieve improved profitability and productivity.

5.3.1 Need for Precision Agriculture

The capability of smart agriculture, with respect to precision farming, to achieve environmental and economic benefits can be visualized through the reduction in the use of water, fertilizers, herbicides, insecticides and fungicides, in addition to the use of farm equipment. Farmers know that their fields have variable inter- and intra-field yields. These variations can be traced back to management practices, properties of the soil and environmental characteristics. Soil characteristics, such as texture, moisture, nutrient status, organic matter, landscape position and environmental characteristics, like weather, water, weeds, insects and diseases, are some of the factors that affect the yield over a farmland. In some fields, within-field variability can be of greater importance than between-field variability. In one field, the best crop growth was observed at the level areas of the field and near waterways. Plants growing on side slopes showed moisture stress at the places where erosion had depleted the topsoil. On seeing this magnitude of variation over their farmlands, most farmers ask how can the problem, that is causing the low yields in parts of the field, be fixed. The management challenge is to optimally manage the areas within the field in order to have different production capabilities, as there is no economically feasible method of fixing the depleted topsoil areas in the field. This does not necessarily mean having the same yield/unit land area in all areas of the field. A farmer's mental information database about how to treat different areas in a field is the result of years of observation and implementation, through trial and error. But today, that level of knowledge about field condition is hard to maintain due to larger farm sizes and changes in farmed areas due to annual shifts in land-leasing arrangements. Precision agriculture paves the way to automating and simplifying the information collection and analysis. It also allows management decisions to be made and quickly implemented on small areas within larger fields.

5.3.2 Technologies Used for Precision Farming, aka Smart Agriculture

It is important for anyone who considers adopting precision farming to be familiar with the modern technological tools available, in order to collect and utilize information effectively. Precision farming requires a vast array of tools, that includes software, hardware and the best management practices, which are described briefly in the following paragraphs.

5.3.2.1 Global Positioning System (GPS) Receivers

The satellites which allow the operation of global positioning systems allow for signal broadcasting, which, in turn, are received by GPS receivers in order to identify the location in question. This information is in real time, which means that information of movement of an object from one position to another position is provided continuously. This allows measurements for soil and crop to be mapped. GPS receivers are either carried to the agricultural field or are fixed. These receivers allow users to return precisely to locations that require sampling or treatment.

5.3.2.2 Remote Sensing

Remote sensing is referred to as data collection from a long distance, mostly wireless. Data that are remotely sensed provide a way of evaluating the health of the crop. Classification of images can easily detect plant stress that is related to nutrients, soil compaction, moisture, or crop diseases. Infrared images, that can be recorded using electronic cameras, correspond to the determination of the health of plant tissue. New image sensors with high spectral resolution are increasing the information collected from satellites. The determination of the location and extent of crop stress can be used in the analysis of results from remotely sensed images. Analysis of such images, used in tandem with field scouting, can help to determine the cause of certain symptoms of crop stress. Thus, a spot treatment plan for crops can be developed and implemented, using the images, to optimize the use of agricultural chemicals.

5.3.2.3 Crop Scouting

Some of the in-season observations of within-field differences in crop condition may include weed patches or fungal infections. Crop tissue nutrient status, water-flooded and water-eroded areas can be spotted using a GPS receiver, so that a location can be linked to observations, making it easier to return to the same location to carry out treatment. Variations in yield maps can then be explained using these observations.

5.3.2.4 Geographic Information System (GIS)

Geographic information system (GIS) involves hardware and software, using location data and feature attributes to produce maps. An important function of an agricultural GIS is to store information in layers. These include yields, remotely sensed data, soil survey maps, reports of crop scouting and nutrient levels in the soil. The geographically referenced data can be displayed in the GIS. In addition to the storage and display of data, the GIS can be used to evaluate present and alternative management by combining and manipulating data layers to produce an analysis of different management scenarios.

5.3.2.5 Information Management

The adoption of precision agriculture requires a combination of information databases and management skills. A farmer is required to have a clear idea of the business objectives and crucial information necessary to make decisions, by leveraging the information obtained. Effective information management is required more than record-keeping analysis tools or GIS.

5.3.2.6 Identifying a Precision Agriculture Service Provider

There exist custom services that help farmers in decision making for effective crop management. They help in offering a variety of services. Custom services can decrease the cost and increase the efficiency of precision agriculture activities by distributing capital costs for specialized equipment over more land and by using the skills of precision agriculture specialists. The most common custom services that are provided by precision agriculture service providers are soil sampling and variable rate applications (using specialized GPS-led equipment) of lime or fertilizer. The equipment that is required for these operations includes a vehicle fitted with a GPS receiver and a soil sampling computer on the field.

5.4 Context of the Work

Figure 5.1 presents the context in which "smartness" can be incorporated into an agriculture-based cyber physical system. To simplify things, cyber physical systems can be broken into two modules: the cyber module, which deals with the data and algorithms required in data processing, and the physical module, which is composed of the actual on-field components that

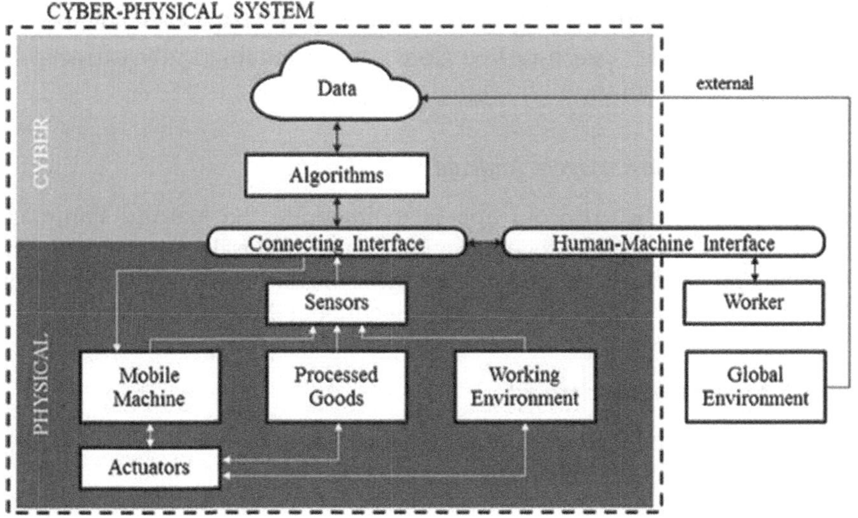

FIGURE 5.1
Context of the work

help in building the entire system. Each component, from the perspective of smart agriculture, is briefly described (Figure 5.1):

The **data**, in the context of smart agriculture, would refer to the images of leaves that are obtained from the crop plants, which are then converted into the required format.

The **algorithm** refers to the process of computing the way in which the classification of a plant type can be done, for example. For this specific example, we will consider the VGG Architecture of ImageNet.

The **connecting interface** refers to the underlying layer between the cyber module and the physical module.

The **human–machine interface** refers to the communication established between the complex system and the user of the system. In this case, the interface is an Android application.

The **sensors** and **actuators** correspond to the sensing of the agricultural environment and acting upon the observations that have been collected.

The **working environment**, in this case, refers to the agricultural field.

The **mobile machine** is connected to the sensors, actuators and the connecting interface. The Android application in the mobile phone and the cloud service are examples of mobile machines.

The **global environment** refers to the cloud [11–13] platform and the database update that takes place over it.

5.4.1 Platforms for Realizing Context

5.4.1.1 Natural Language Processing

This is achieved with the help of the Google Text-To-Speech Converter, as well as the Google Speech-to-Text Converter. It automatically extracts the required words from the audio signals (Figure 5.2).

5.4.1.2 Application Service – Android

The purpose of the Android app is to integrate the Natural Language Processing (NLP) services together with the image classification model. The Android app can be developed using either Android Studio or the MIT App Inventor. In this scenario, we use the conventional Android Studio IDE.

5.4.1.3 Neural Network Model

The correct choice for the neural network needs to be made. There are two possible choices for this situation:

1. In order for the model to be trained from scratch, it typically requires a Python script to run. By convention, Keras deep-learning model is the preferred form because of its seamless integration. Hence, we choose the Keras model for the Python script.

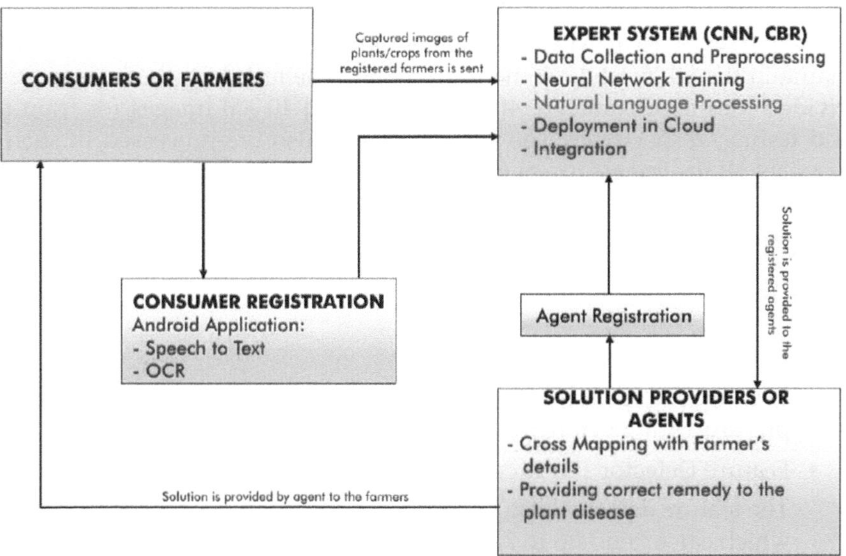

FIGURE 5.2
The block diagram detailing the process

2. There is another case, where the trained model needs to be deployed into the Android app. Buta drawback here is that Keras is not supported by Android. As a result, the TensorFlow deep-learning model can be used to deploy the model to Android.

5.4.1.4 Cloud Service

Cloud Service serves the purposes of both using database connectivity operations from the application as well as deploying the model in the chosen cloud service, such that the appropriate class label can be obtained. For this purpose, we go for Amazon Web Services (for deploying the model to cloud [5–9]) and the Sheets API Service–Google Drive (for all the database operations).

5.5 Prototype Implementation

5.5.1 Implementation of Phase 1 and Phase 2 (Data Collection, Pre-Processing and CNN)

5.5.1.1 The Dataset

The dataset consists of approximately 50,000 images [10]. All of the images were obtained from an open-source data collection of images, the PlantVillage

Dataset. It has a total of 14 different crops along with 38 different diseases; some of the classes also include images of healthy plants as well. In order to maintain the 80:20 ratio for the training and the test data, the images were divided into approximately 40,000 images and 10,000 images for training and testing, respectively. The input image is also pre-processed in such a way that all the images are of fixed dimensions and size.

5.5.1.2 Step 1 – Convolution

There are three components involved here:

- Input Image

  The input image is the image being detected. It undergoes a phase of pre-processing before it is passed to the next stage.
- Feature Detector

  The feature detector, also known as a filter or kernel, is a 3 ×3 matrix, which can extend up to 7 × 7.
- Feature Map

  The input image matrix is multiplied by the feature detector matrix to produce the feature map. The aim of this step is to reduce the size of the image and to make processing faster and easier. Some of the features of the image are lost in this step, although the important features are retained. These features are the ones that are unique for identifying that specific object.

5.5.1.3 Step 2 – Using a Rectified Linear Unit

The rectifier function is applied to increase non-linearity for the neural network. Non-linearity exists in images which are made of different objects. Without applying this function, the problem of image classification would be linear.

5.5.1.4 Step 3 –Max Pooling

This concept, where the ability of a neural network to detect specific features of an image is not affected by the location of an object in that image, is called spatial invariance. Pooling helps in feature detection for the Convolutional Neural Network (CNN), with respect to different angles and light exposure of the image. Max pooling is where the largest value is picked among a 2 × 2 matrix placed on a feature map. The matrix is moved sideways to cover the entire feature map, repeating the same process. The resultant values are called the pooled feature map. It helps in maintaining the essential features of the image, while reducing its size. This helps to reduce overfitting, which happens when irrelevant information is stored in the neural network.

5.5.1.5 Step 4 –Flattening

The next step is to flatten the pooled feature map, which involves transforming it to a single column and inputted to a neural network, which is used for further processing. This is then passed to a neural network. This step is made up of the input layer, the fully connected layer and the output layer. The predicted classes are obtained in the output layer. The prediction error is also calculated in the final step. To improve accuracy, the error is propagated backwards through the network.

5.5.1.6 Training the Model – Using Keras and VGG

The training for the model would require the Keras library and its "layers" module. As mentioned above, there are five components that are required for creating the neural network:

1. 'Sequential', which is used in neural network initialization.
2. 'Convolution2D', which is used to create the convolutional network with respect to images.
3. 'MaxPooling2D', in order to add the max pooling layers.
4. 'Flatten', which is used to convert the featured map (which is pooled) to one single column, which is then passed on to the layer, which is fully connected.
5. 'Dense', which can be used to add the fully connected layer to the respective neural network.

The first step would be to choose an architecture for ImageNet. This is done so that the pre-trained weights can be assigned to the nodes in the network. We choose VGG16 for this purpose, as it provides greater accuracy in terms of implementation. We also set the input shape that is to be inputted to the neural network. Here, the size is specified as a 224 × 224-dimension image.

The next step would be to create a model for the neural network. First, we configure the model to be sequential. This happens in response to adding the vgg weights to the network. Then, we add the input and the hidden layers, specifying the activation function as well. In order to maintain the accuracy of the model, we randomly dropout some of the input images that are given to the network. This is called the dropout ratio. It is set to a typical value of 0.5. The final step would be to add the output layer. We specify it as 38 nodes because the number of class labels that the neural network has is 38 (corresponding to the 38 different plant diseases).

The next step would be to compile the model. The pre-requisite for this step would be to set up the training and the testing paths for the images. We also set the number of epochs, the steps per epoch, the paths to the training and the testing data.

We then test the model to see if it classifies the given input image correctly. We first re-size the input image. Then, using the model, we find out the predicted class label, which can be cross mapped to identify the exact plant disease.

5.5.2 Implementation of Phase 3 (Natural Language Processing)

5.5.2.1 Optical Character Recognition

The process of detecting text in images and video streams, and recognizing the text contained therein is Optical Character Recognition (OCR). After detection, the recognizer determines the actual text in each block and segments it into lines and words. The Text API detects text in many languages, as it provides custom support for many Latin-based languages, in real time, on the device. The objective of creating this OCR-based application is that the user need not type in his/her entire address details manually, as this would consume a brief amount of time. But, with the help of OCR, this can be greatly reduced, and the time of processing is also reduced to a large extent.

The text recognizer segments text into blocks, lines and words. Roughly speaking:

- A Block is a contiguous set of text lines, such as a paragraph or a column.
- A Line is a contiguous set of words on the same vertical axis.
- A Word is a contiguous set of alphanumeric characters on the same vertical axis.

So, the text recognizer would split up the entire text to be recognized in a hierarchical fashion, split it up into the simplest form possible and then recognize each of those individual words. That is the main purpose of using the Text API for an Android application. The main requirements for this application would be the user's permission to allow the application to access the camera.

5.5.3 Steps Involved in Creating OCR

The first step would be to add the Text API as a dependency to the android application project. To use this library, we may need to update the installed version of Google Repository in SDK tools. So, we need to make sure that the version of Google Repository is up to date, at least version 26. We would have to declare the text recognizer object in order to start with the recognition of text from the image obtained from the camera. We will now need to create the camera source function in order to capture the image.

In the camera source object, we set the following parameters:

1. The side facing the camera – whether it is front facing or backwards facing.
2. The requested preview size – the size we would like the camera to occupy on the screen.
3. The auto focus option – set it to 'true'to automatically focus on the image.
4. Set the frames per second (fps) value – 2.0f (a float value) by default.

Finally, we build the camera source object. The next important step would be to set the permissions of the camera at run time, i.e., permission obtained from the user. We check whether the permission has been granted by the user or not. If not, then we set the respective camera permissions for the user and we also start the camera function. The final step in this process is to create the text recognizer function that recognizes the respective text obtained from the image.

5.5.3.1 Speech-To-Text Conversion

The next module that is present is the speech-to-text conversion application. This was specifically developed, keeping in mind that farmers may not be able to type text onto a smartphone. Hence, considering this factor, it would be easier for them to give a voice input. The Android application would convert the voice into the corresponding text, which would be stored.

5.5.3.2 Steps Involved in Creating Speech-to-Test

The basic layout would be a voice input module, which listens to user input, and the text view to display the text corresponding to the voice. This has to be set up in the activity_main.xml file. The next step would be to configure the MainActivity.java file that essentially contains the required permissions for the application and the configurations for the voice input and the text view. Here, we create the object references for the voice input and the text view buttons that we have declared in the xml file. We also configure the button OnClick method, which, in turn, calls the askSpeechInput() method.

We set the request code for speech input as 100. In this way, we make sure that the Android application has received permission for recording audio. This additional check is made inside the next corresponding method. In this function, we check the input only for the correct request code. Here, we set the value of the result obtained to an array list and display this text in the corresponding text view on the screen.

5.5.4 Implementation of Phase 4 – Deploying the Model as a TensorFlow Graph

Now, we have the trained model in Keras. But, in order to deploy the neural network to an Android application, we would have to convert the Keras model to a TensorFlow graph, because Keras is not supported by Android. Hence, TensorFlow could be used for this process of transformation.

5.5.4.1 Transformation to TensorFlow

The final model for Keras is stored in the form of a h5 model. TensorFlow, by default, stores all the weights and nodes in the form of a protobuf file. The protobuf file is essentially a frozen graph, where all the weights are frozen. We also need to convert the h5 model to a .pbtxt file as well. The process is done by using a Python script to convert it to the required form

The export_model has four parameters:

- tf.train.Saver()–it saves and restores the variables that are being used in the neural network model.

It adds ops to save and restore variables between checkpoints.

- model – the Keras model itself.
- input_node_names – the names of the input nodes in the neural network.
- output_node_names – the names of the output nodes in the neural network.

With these parameters being passed to the function, the Python script would carry out the following:

1. From the TensorFlow backend, which represents the Keras model, the graph for computation needs to be obtained. The function would create a new computational graph by pruning out all the unnecessary subgraphs that are not required for obtaining the output.
2. The next step would be to write the computational graph using the tf.train.write_graph function. It takes the current TensorFlow/Keras session as the input. We can also set the resulting graph in a specified path. The output would be a serialized GraphDef proto.
3. Following that, we need to save the current session as an intermediate checkpoint. It is saved as .chkp format file.
4. This is the step where we freeze the graph. Then, the graph will be converted to a GraphDef protocol buffer. After that, as mentioned

earlier, some of the subgraphs, which are not necessary for output computation, are pruned out.

5. In order to use the graph, we deserialize it from the .pb format, using the ParseFromString() function.
6. We also use the optimize_for_inference module, which takes a frozen binary GraphDef file as input.

It outputs the GraphDef file which we can use for inference.

5.5.4.2 Deploying the Transformed Model

Now that we have converted the .h5 saved model in the form of a frozen graph, the next step would be to deploy this model in Android. We create a Kotlin class for this purpose.

1. This Kotlin class would implement a camera interface, where the user clicks a picture of a plant. The TensorFlow image classifier would resize the complete image into the required input size for the frozen graph. The result is displayed, based on the percentage of confidence of the classified leaf type.
2. To start things off, we set up a list of constant objects, which contain the following input parameters to the TensorFlow image classifier:
   - TAG – the name of the MainActivity
   - INPUT_WIDTH, INPUT_HEIGHT – the required dimensions for the input, to be passed to the classifier.
   - IMAGE_MEAN, IMAGE_STD – the mean and the standard deviation range for the input image.
   - INPUT_NAME – the name of the input node of the frozen graph.
   - OUTPUT_NAME – the name of the output node of the frozen graph.
   - MODEL_FILE – the uploaded frozen graph.
   - LABEL_FILE – this contains the list of all the class labels that are present for our model.
3. We represent, as an example, the "initialiseClassifier()" method, which contains the following:
   - A tensorflowimageclassifier.create() function, which contains the list of constant objects as input parameters:
   - TensorFlowImageClassifier.create(assets, MODEL_FILE, LABEL_FILE,INPUT_WIDTH, INPUT_HEIGHT, IMAGE_MEAN, IMAGE_STD, INPUT_NAME,OUTPUT_NAME)

The input argument "assets" represents the resource folder which contains the uploaded frozen graph.

4. After the function has correctly classified the input image, we instantiate the "onCaptured" method, which does the following:
   a. It contains a camerakit object as the input, which contains the captured input image.
   b. It creates a scaled bitmap for the captured input image. This would resize the input image to the size required for the frozen graph.
   c. We display the captured image, as well, to the screen.
   d. We internally invoke another method called "showRecognizedResult" for the resized input image.
5. In the "showRecognizedResult" method, we obtain the class label and the confidence level (i.e., the percentage of accuracy) for the given input image. Both of them are stored in a "results" object.
6. We display the final result on the activity screen. We simultaneously update them in the sheet database as well.

5.5.5 Implementation of Phase 5 – Integration of Cloud Database

All the user details provided need to be updated in a central database. For this purpose, we use the Sheets API from Google Drive.

5.5.5.1 Required Dependencies

In order to access the Sheets API from Google Drive, we require a list of dependent libraries that need to be added:

1. The Volley Library, which would help in dealing with the REST request and responses.
2. The Google Play Services package, for basic Google Services.
3. The Google API Client package, for authentication.
4. The Google API Services–Sheets package, for accessing Google Sheets.

5.5.5.2 Accessing Google Shifts

In order to access this sheet, the first step would be to create a Google Apps Script. Google Apps Script allows you to access any Google-based application from any external application. A JavaScript function needs to be created in order to establish the connectivity between the Android application and the Google Sheets. We also use the Sheets API service. For instance, the sheets.getRange() function would allow us to access the complete Google Sheets, mentioning the starting point of the row, the starting point of the column, the range of the row and the range of the column. After this step, we iterate through the Google Sheets, store each entry in the table in a list as a record. The final step would be to convert this to a JSON object using the JSON.stringify method. We finally return this object to the calling request URL. In order to generate the URL, we need to deploy this Apps Script as a webjob. The required permissions for sharing the file need to be provided. If the steps are followed correctly, a URL will be generated.

5.5.5.3 Implementation in Android

In order to use the URL for accessing the sheet, the Volley library from Android Studio can be used.

Here is the list of steps required to create the application.

1. Create a stringRequest object. The object contains two parameters, the first parameter denoting whether the request is a GET request or a POST request, and the second parameter denoting the URL that needs to be accessed.
2. Create the request parameters to be added. A Hashmap needs to be created and added as entries to the parameter to the object.
3. We give the request a certain amount of time, such that the function can try to establish the connection. Now, the application would be able to successfully process the request and access the GoogleSheets.

5.6 Observation

The approach that needs to be adopted by the policy makers to promote smart agriculture in the context of precision farming includes to:

- Promote this smart technology for specific farmers who have the capability to bear the capital investment risks.
- Identify areas where organic farming can be incorporated.
- Encourage farmers to adopt protocols related to water management.
- Promote the usage of irrigation systems at the micro level and of the ways in which water can be saved.
- Study spatial and temporal parameters using data obtained at the field level.
- Develop a policy for the transfer of efficient technology to the farmers.

- Provide complete back-up support to the farmers, with respect to the technical side of smart farming, which can be incorporated and replicated on a large scale.
- Provide policy support for cooperative groups.

5.7 Discussion and Future Scope

Precision agriculture helps farmers to use crop inputs more effectively, which includes pesticides, tillage, fertilizers and irrigation water. This targets greater crop yield with reduced risk of environmental pollution. However, the cost–benefits of managing precision agriculture are difficult to determine. In the present time, most of the technologies that are used are still in their infancy, so that pricing them would be difficult. Precision agriculture takes care of both environmental and economic issues of production agriculture. The precision agriculture success depends mainly on the knowledge needed to guide new technologies to farmers. This is the way in which "smartness" can be incorporated into agriculture. From the perspective of the future, several smart agriculture technologies, that have already been described here, will soon be incorporated into existing field-testing techniques.

References

1. http://www.fao.org/family-farming/detail/en/c/897026/.
2. https://www.sciencedirect.com/science/article/pii/B9780128038017000250.
3. https://www.researchgate.net/publication/311996916_Agriculture_Cyber-Physical_Systems.
4. www.tlc.unipr.it/ferrari/Publications/Journals/FrFe_AST18.pdf.
5. GR Karpagam, J Parkavi 2011. "Setting up of an open source based private cloud", *IJCSI International Journal of Computer Science Issues.*
6. J Uma Maheswari, GR Karpagam 2014. "Ontology based comprehensive architecture for service discovery in emergency cloud", *International Journal of Engineering and Technology (IJET)* 6(1), 242–251.
7. GR Karpagam, J Uma Maheswari 2015. "Cloud based comprehensive architecture (CBCA) for QoS aware semantic web service discovery in emergency management system", *International Journal of Applied Engineering Research* 10(23), 43763–43771.
8. J Uma Maheswari, GR Karpagam, A Bharathi 2014. "Self-organizing agent based framework for service discovery in cloud", *Proceedings of International Conference on Advances in Communication, Network and Computing*, Elsevier, pp. 877–888.

9. S Subramanian, NG Krishna, KM Kumar, P Sreesh, GR Karpagam. 2012. "An adaptive algorithm for dynamic priority based virtual machine scheduling in cloud", *International Journal of Computer Science Issues (IJCSI)* 9(6), 397.
10. B Vinoth Kumar, GR Karpagam Manavalan 2011. "An empirical analysis of requantization errors for recompressed JPEG images", *International Journal of Engineering, Science and Technology* 3(12), 8519–8527.
11. S Vijayalakshmi, GR Karpagam 2018. "Secure online voting system in cloud", *Electronic Government, an International Journal* 14(3), 276–286.
12. GR Karpagam, S Vijayalakshmi 2018. "Authentication as a service in cloud from a fuzzy perspective", *International Journal of Enterprise Network Management* 9(3–4), 352–362.
13. S Vijayalakshmi, GR Karpagam 2016. "An agent based online voting system in cloud using blind signature", *Asian Journal of Information Technology* 15(19), 3826–3834.

6

Building Cyber Physical Systems in the Context of Smart Cities

Malligeswaran Jagannathan

CONTENTS

Organisation of the Chapter

Section 1 presents the terms and terminologies for the user to understand the chapter. **Section 2** provides a high-level introduction to the importance of cyber physical systems in the context of smart cities. **Section 3** highlights a number of key physical processes that are critical for driving smart city maturity. **Section 4** covers the major requirements for achieving an effective deployment of cyber physical systems. **Section 5** details the benefits of deploying a cyber physical system to support smart city initiatives. **Section 6** details the current challenges in deploying cyber physical systems for smart city initiatives, specifically in developing economies. **Section 7** summarizes the expected evolution of cyber physical systems with the developing technology, that brings with it its unique challenges and opportunities.

6.1 Terms and Terminologies

- The Internet of Things (IOT) – The Internet of Things is a system of interrelated computing devices, mechanical and digital machines, objects, animals or people that are provided with unique identifiers and the ability to transfer data over a network without requiring human-to-human or human-to-computer interaction.
- Artificial Intelligence (AI) – In computer science, artificial intelligence, sometimes called machine intelligence, is intelligence demonstrated by machines, in contrast to the natural intelligence displayed by humans.
- Machine Learning – Machine learning is the scientific study of algorithms and statistical models that computer systems use to perform a specific task, without using explicit instructions, relying instead on patterns and inference. It is seen as a subset of artificial intelligence.
- Cyber Physical Systems (CPS) – A cyber physical system is a system in which a mechanism is controlled or monitored by computer-based algorithms.
- Pervasive Sensing – This is an approach to collecting data by adding wireless sensors, rather than connecting devices using wires and cables, to take advantage of the lower cost, ease of installation and greater data availability.
- Virtualization – Virtualization is a technology that lets you create useful IT services, using resources that are traditionally bound to hardware. It allows you to use a physical machine's full capacity by distributing its capabilities among many users or environments.

- Data Cleansing – This is the process of detecting, correcting and negating incorrect records from a dataset, table or database, so that only the useful data are retained for analytics.

6.2 Introduction

In the era where everyone has access to the Internet and technology, terms like the Internet of Things (IOT), digital transformation, big data, analytics, artificial intelligence and machine learning have increasingly found their place in everyday communication and activities.

To realize these discussions, there are two approaches:

- Top-down approach
- Bottom-up approach

In the top-down approach, technologies and models of deployment tend to expand from the broader high-level computing analytical platform to devices or sensors as low-level physical entities.

On the other hand, with the bottom-up approach, low-level devices or sensors take precedence over connectivity to high-level computing analytical platforms.

Each of these approaches has its own advantages, depending on the vertical pillars of physical entities where CPS is deployed and major trait that qualifies the urban core of smart cities.

With the fast-evolving technology and the tremendous growth in thin adaptation to the technology in everyday life, even by people at the grassroots level, it has become inevitable that cyber physical systems are built to support every aspect of life.

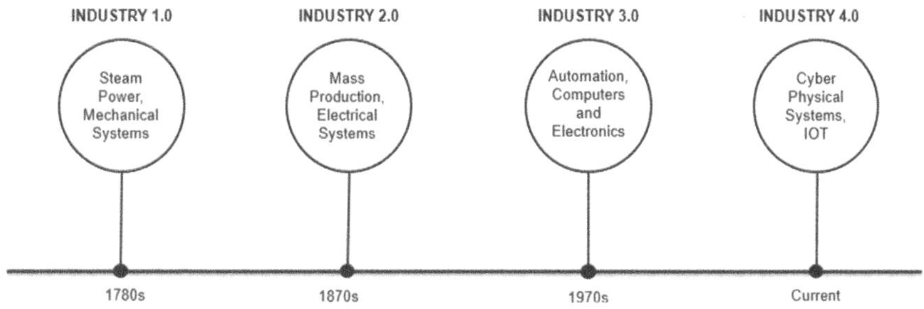

FIGURE 6.1
Industrial revolutions

Being one of the core components of the fourth industrial revolution, Industry 4.0, CPS converges the integration of social computing and Industrial IOT into a single platform to leverage interaction, cooperation and learning from multiple layers of physical and logical systems (Figure 6.1).

The predominant shift towards widely used physical components, empowered with cyber capabilities, and the amount of data available over its integration with other systems, shows the path ahead in realizing more smart cities in our neighbourhood (Figure 6.2 and Table 6.1).

6.3 Fuel for Cyber Physical Systems

Before starting any discussion on the deployment of CPS, it is vital to map the various physical processes and how these processes lineup to achieve maturity of the ultimate smart city (Figure 6.3).

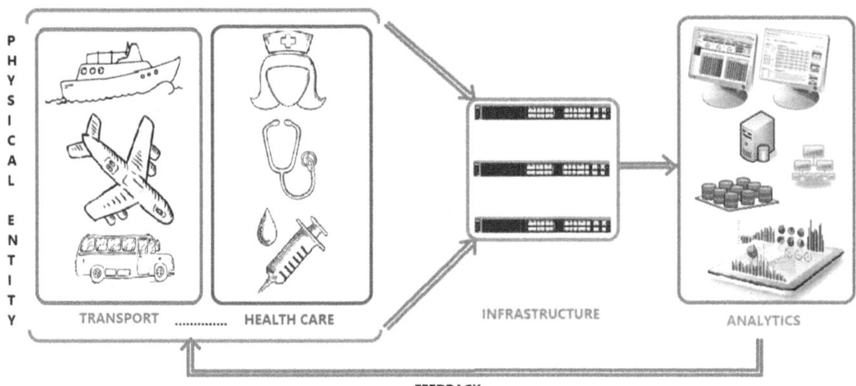

FIGURE 6.2
CPS – typical architecture

TABLE 6.1

Abbreviations

- CPS – Cyber Physical System
- ETL – Extraction, Transformation and Loading
- ISA – International Standards on Auditing
- IOT – Internet of Things
- IT – Information Technology
- MB – Megabytes
- NERC – North American Electricity Reliability Corporation
- NIST – National Institute of Standards and Technology
- ROI – Return on Investment

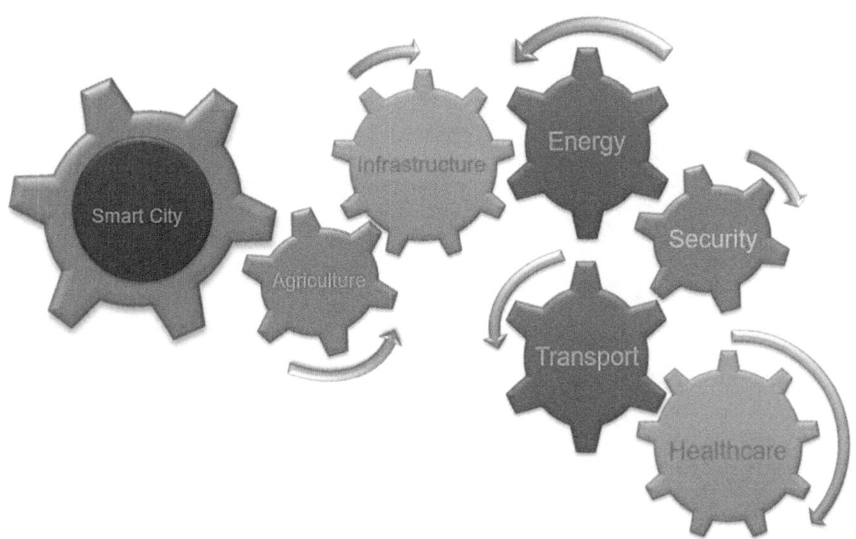

FIGURE 6.3
Key physical processes

For an effective deployment of CPS, a perfect integration is essential between physical processes and a computation analytical platform that provides modelling, design and analysis to accommodate physical dynamics.

Recent research advances have mainly focused on the following respects: energy management, network security, data transmission and management, model-based design, control technique, system resource allocation and applications.

Wireless sensing, CPS, IOT, analytics, artificial intelligence and machine learning have emerged, which attracted and engaged many researchers and organizations. In recent years, higher rates of adoption have stimulated growth in these emerging fields, and these achievements have fuelled the development of CPS to a great extent.

Important features that need to be considered in deploying a CPS environment are described in Figure 6.4.

The success and benefits of a CPS lie in finding a balance between the four features of security, usability, scalability and flexibility. When a design tends to provide more of one feature, the design loses more of the other three features, so that finding a balance between these features in deployment is an absolute necessity.

Several studies have been carried out in recent times to deploy CPS at all fronts of physical processes, but the processing and analysis of captured data have proved to be the slower process and represents the greater challenge, for various reasons, specifically in developing countries.

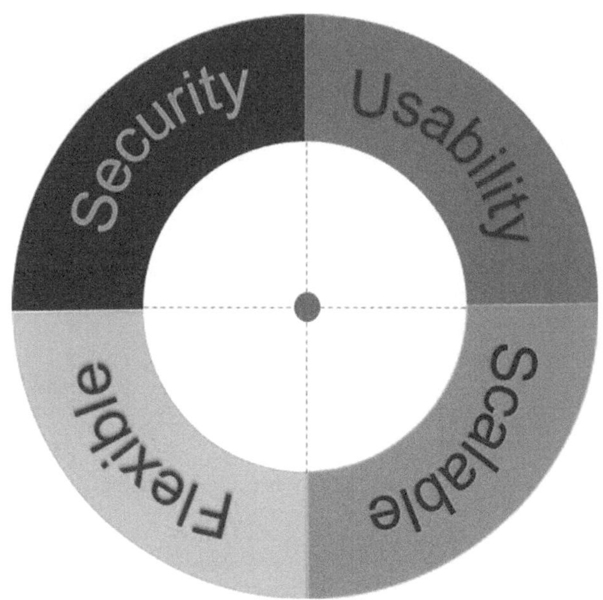

FIGURE 6.4
Aspects of CPS deployment

6.4 CPS as the Backbone of Smart Cities

6.4.1 Infrastructure

One of the main requirements for any approach to implementing CPS is the infrastructure required on both the data source and connectivity layers and the analytical layer, where the data are processed and inference can be achieved.

A data source should comprise intelligent field devices, pervasive sensing and acceptable communication protocols, without compromising the requirements of ease of deployment, accuracy, speed in communication and relevance.

Data sources:

i. Physical sensors
ii. Fixed sensors
iii. Mobile sensors
iv. Wireless sensing

v. In-built smartsensors
vi. Online data sources

6.4.2 Data Handling

6.4.2.1 Data Collection

- The definition of what data need to be collected, at what intervals and how to collect that data comes much earlier in deployment, as the whole analytic and inference decisions come from these data.
- Data collection can be quantitative and qualitative.
- Quantitative data collection methods rely on random and structured data collection, that are used in producing results that are easy to summarize, compare and generalize. Qualitative data collection methods rely on providing data that are both useful and accurate in order to understand the process and the results in in-depth, documentable actions.

6.4.2.2 Data Storage

- Data storage is the process of consolidating data from different sources into a centralized location and in a format acceptable and useable for further data management and analytics.
- As the model adopts scalability, it becomes vital to maintain the simplicity and unification of data storage so that retrieval of data, when needed, is easier.
- As historical data provide information on process dynamics, effective handling of bulk data and identification becomes important for analytical applications.

The collected data might have information that is sensitive and proprietary, so care needs to be taken in handling such information securely and effectively. With the latest technologies and the amount of data to be handled and processed, virtualization and Big Data historians extended from the IT environment, where data handling has been advanced and effective, can be adopted (Figure 6.5).

6.4.2.3 Data Visualization

i. In a complicated CPS deployment, like the one in the context of smart cities, data come from various sources and in multiple formats. Visualization of the data collected provides insights on hidden data.

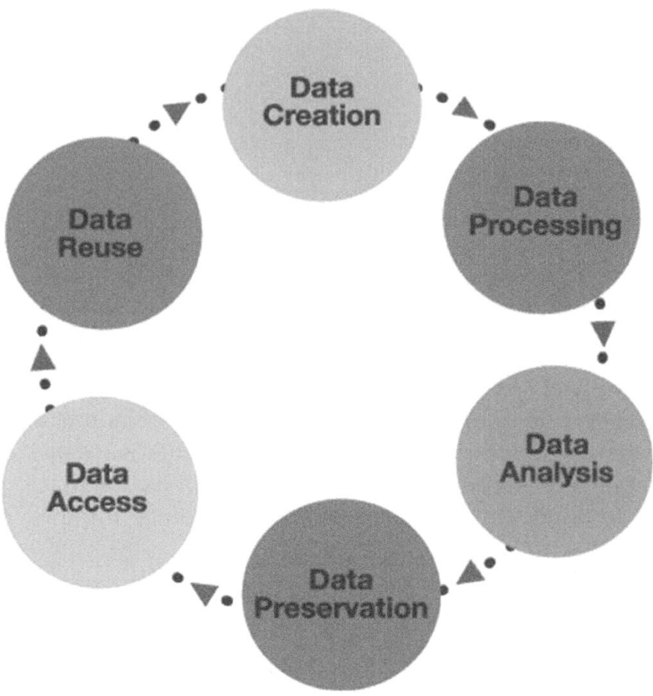

FIGURE 6.5
Data handling cycle

ii. Data visualization offers a swift, intuitive and simple way of analyzing the data.
iii. It is estimated that approximately 1.7 MB of data every second will be generated per human on the planet by end of 2020. With the overwhelming level of data and insights available in today's digital world, handling, interpreting and presenting this rich wealth of insight proves to be a real challenge. Breaking down the data for the most focused, logical and digestible visualization is possible by:
    - Discovering which data are available and which data are most valuable.
    - Understanding from where data comes and how to access it.
    - Keeping the data protected, and presenting the most valuable insights.

6.4.2.4 Data Cleansing

- Data cleansing or data cleaning is the process of identifying and removing or correcting inaccurate records from a dataset.

- There are several tools and methods available for cleansing the data collected. Cleansing does not strictly mean deleting irrelevant data but should also facilitate recognizing patterns in the data for unfinished, unreliable, inaccurate or non-relevant parts of the data and then restoring, remodeling or removing the dirty or crude data.
- In fact, with a properly cleansed dataset, even simple algorithms can gain impressive insights from the data. Different types of data will need different types of cleansing.
- Incorrect or inconsistent data lead to false conclusions. As a consequence, how well you cleanse and understand the data has a significant impact on the quality of the results.
- Not cleansing the data can lead to a range of problems, including linking errors, model misconfiguration, errors in variable estimates and incorrect analytics, leading to making false conclusions.

6.4.2.5 **Data Integration**

Data integration is the combination of technical and business processes used to combine data from different sources into meaningful and valuable information.

The process of data integration is about taking data from many sources and combining them to provide a unified view of the data for business intelligence.

Data integration in general is defined as a three-step process:

i. Extraction:

   Extraction involves connecting to the source systems, and both selecting and collecting the necessary data needed for analytical processing within the data warehouse or data mart. Usually, data are consolidated from numerous, disparate source systems, that may store the data in a different format. Thus, the extraction process must convert the data into a format suitable for transformation processing. The complexity of the extraction process may vary, and it depends on the type and amount of source data.

ii. Transformation:

   The transformation step of a data integration process involves execution of a series of rules or functions to the extracted data to convert it to a standard format. It includes validation of records and their rejection, if they are not acceptable. The most common processes used for transformation are conversion, clearing the duplicates, standardizing, filtering, sorting, translating and looking up or verifying whether the data sources are inconsistent.

iii. Loading:

The load is the last step of the "extract, transform and load" (ETL) process and involves importing extracted and transformed data into a target database or data warehouse. Some load processes physically insert each record as a new row into the table of the target warehouse. The bulkload routine maybe faster for loads of large amounts of data but does not allow for integrity checks upon loading of each individual record.

6.4.2.6 Data Interpretation

With the sheer amount of data available in this digital age, the ability to analyze complex data, producing actionable insights and adapting to new market needs at the speed of our thoughts, is vital. Dashboards are the tools capable of providing the user interface to visualize key performance indicators for both qualitative and quantitative data analysis.

Data are very likely to arrive from multiple sources and tend to enter the analysis process with haphazard ordering. The nature and results from interpretation will vary for different processes.

Key attributes of data in today's physical environment are volume, speed, variety and value. Interpreting such a magnitude of data needs effective visualization, that enables rapid exploration of data and identification of relationships between those data parameters.

When interpreting data, caution should be taken to factor in any bias in the data for various reasons that could lead to inaccurate interpretations. Fault tolerance levels in each implementation will differ, depending on the criticality of the physical systems.

6.5 Opportunities from CPS in Building SmartCities

6.5.1 Performance

Most smart city projects are based on sensor-centred collection and analysis of data that provide a cost-effective and innovative solution to the growing challenges.

Smart city projects go far beyond simply providing connectivity for all, or offering access to information on public services, such as transport. Smart city initiatives are digitally transforming public services, completely altering the way in which the environment is constructed and managed, and the way we interact and live within these environments.

The smart city, as a holistic vision, is about more than just technology. It is about thinking smarter to solve real-life human problems.

Every city is unique in terms of population, climate, traffic patterns, and judicial laws and so, when supporting basic infrastructure for a smartcity, these details should be considered.

With advances in technology, more smart products and services are available and implemented. These products and services, when highly interconnected, provide expected cost- and energy-efficient, tangential, meaningful, user-centric and sustainable realization of smartcities.

With a distributed variety of systems interconnected, identifying the weak link early in the process is crucial. Identifying such links and applying the inferences to improve their performance to support realization of smartcity initiatives is vital.

Periodic offline data analysis helps in identifying any change in the process dynamics that call for changes in the models, that can mimic the real-time process.

The outcome of such offline data analysis is to be taken and embedded in the model for online analysis and monitored continuously for validation.

6.5.2 Economical

For any initiative, justifying the economic viability and return-over-investment (ROI) is key to proceeding. In general, projects are implemented in phases, and achieving these key values in a progressive methodology needs to be considered (Figure 6.5).

Smart city initiatives start with mobilizing the right resources and the right funding at the right time.

With fast-evolving technology, components that build up the CPS are bound to become outdated very soon. Revalidation and progressive development in the design of models and components are important.

The close integration between physical systems and analytic platforms provides the complete control needed to make any economic decision and to predict the value proposition from such initiatives and, in turn, to drive ROI.

6.5.3 Scalable

It is never a one-time job to implement a system of such magnitude and impact. The model should show inherent flexibility in exploiting digital technologies and a supporting capability to create a robust new digital business model.

Apart from technological advances in the way CPS are deployed, there is a lot of development in the foundational solutions that needs to be considered when it comes to adoption to scalability.

When a deployment is designed for integration with complex multi-layer physical systems, scalability is inherent in design and is not only for expanding to other locations but also to other processes.

The IT world is highly advanced in adopting the scalability and, with the latest cloud-based and virtual environments, any degree of scalability is possible.

6.6 Challenges in Adapting CPS to Build SmartCities

6.6.1 Security

Data security has gained significant importance in recent years, with the free knowledge available over the Internet and more tech geeks exploring how to exploit it. CPS deployment and analytics is data driven and hence ensuring data security is key and challenging.

Investment in technologies such as 5G, big data and IoT can be successful only when the technologies deployed are genuinely able to improve the quality of life for a city's inhabitants, making every-day processes more streamlined, efficient and sustainable. The development of a trustworthy smartcity requires a deeper understanding of potential impacts resulting from successful cyberattacks. Estimating feasible attack impact requires an evaluation of the dependency of the physical process on its cyberinfrastructure and its ability to tolerate potential failures.

A further exploration of the cyber physical relationships within the process and a specific review of possible attack vectors is necessary to determine the adequacy of cybersecurity efforts.

With the advantages of CPS adaptability to different system landscapes comes the risk of larger attack surface. Unlike the IT infrastructure and strategies to safeguard cyberthreats, CPS security needs an understanding of the security features and requirements for specific processes.

System security design should accommodate:

i. Prevention:

   We believe that the major research challenge for preventing the compromise of physical systems is to identify ways in which asset owners and vendors of physical systems will be motivated to follow the best security practices.

   There are several industry standards based in the sectors that can be adopted, like NERC for the electricity sector, NIST for general IT in special publication 800-53, and ISA-SP 99 for manufacturing and general industrial controls.

ii. Detection and Recovery:

   Since we can never rule out successful attacks, security engineering has recognized the importance of detection and response to attacks.

Whereas traditional intrusion detection systems look at network or computer system traces, control system scan provide a paradigm shift for intrusion detection by monitoring the physical systems for abnormalities, so that we may be able to detect attacks that are undetectable from the IT side.

iii. Resilience:

There are several security design principles that can be useful for designing cyber physical systems that can survive attacks, like redundancy, diversity, principle of the least privilege, separation of privilege, etc.,

iv. Deterrence:

It will be essential to enforce law, legalization, and international collaboration for tracking crimes beyond physical boundaries.

6.6.2 Physical System Dynamics

Changes in physical systems vary, based on the complexity and distribution of integrated subsystems. The more complex and distributed a system is, the more changes in dynamics can be expected.

In most scenarios, the changes in physical dynamics are helpful in identifying new events not available in the data collected, building a robust model, provided that a mechanism for detecting such changes is in place and a feedback mechanism to validate the changes is available.

6.6.3 Management

Technology innovation is the enabler that improves the possibilities and efficiencies of each smart city project. Each new technology brings with it an immense pool of new possibilities. Since every city has its own culture, infrastructure and funding policies, technology adoption can vary in diverse ways. However, that means it is not always possible to rely on other proven smart city projects to act as a blueprint for success.

Scope should be embedded for easy and remote upgrading as the technologies evolve. A contingency plan should be devised and implemented for any unexpected rollback requirements.

6.7 Conclusion

In the past few years, this emerging domain for CPS has been attracting significant interest and this will continue for years to come. Despite rapid evolution, smart cities are still facing new difficulties and severe challenges.

- Technology challenges with coverage and capacity.
- Digital security.
- Legislation and policies.
- Lack of confidence and lack of clarity around benefits.
- Funding and business models.
- Interoperability.
- Existing infrastructure to support cyber physical systems.

The definitions of CPS and IOT are converging and might result in a hybrid system of tightly integrated digital, analogue, physical and human factors. This can bring together those sectors that are isolated to evolve as a shared, innovative and redefined system.

The functional requirements, usage and outcome of a CPS implementation should be considered when designing, building and validating these systems.

A common unified structure, with scalable technology and unified components, should be considered to enable open and reliable architecture.

At each stage of design, development and deployment, provision should be incorporated to verify, re-evaluate and re-scope the objectives, so that a tightly integrated physical and logical state can be achieved.

Bibliography

1. Industrial Revolutions [Online]. Available: https://www.istockphoto.com/.
2. Industry 4.0 [Online]. Available: https://nordigi.no/index.php/en.
3. NIST Cyber-Physical Systems for Global Cities [Online]. Available: https://www.nist.gov/programs-projects/cyber-physical-systems-global-cities.
4. Key Physical Processes Driving Cyber Physical Systems Impacting Smart Cities [Online]. Available: http://cyberphysicalsystems.org/.

7

Cyber Physical Systems for Disaster Response Networks

Conceptual Proposal using Never Ceasing Network

I. Devi, G.R. Karpagam and J. Uma Maheswari

CONTENTS

Organization of the Chapter

Section 1 presents the terms and terminologies for the user to understand the chapter. **Section 2** introduces disaster response networks. **Section 3** gives the research background and the aim of the present work. **Section 4** shows a proposed architecture of the Never-Ceasing Network for disaster response networks. **Section 5** and **Section 6** show the system prototype and tools. **Section 7** depicts the disaster response networks. **Section 8** shows the scheduling algorithms while **Section 8.1** presents the precedence scheduling algorithm. **Section 9** gives a comparative analysis. Finally, **Section 10** concludes this chapter and presents future work.

7.1 Terms and Terminologies

Cyber Physical Systems (CPS) – An incorporation of network, computation and physical processes.

Virtual Machine (VM) – A virtual machine is a computer file, usually called an image that acts like a real computer. In other words, VM is the construct of a computer within a computer.

Internet of Things (IoT) – IoT is the system of organized things/devices that are embedded with sensors, software, network connectivity and essential electronics, which permits it to gather and exchange data.

Never-Ceasing Network (NCN) – The Never-Ceasing Network is a novel network paradigm to provide network capability in both normal and disaster-based environments.

A Cognitive Radio (CR) – The cognitive radio makes observations of the environment, makes decisions and is reconfigured, as a result of the observations, learning from the experience.

A Cognitive Radio Network (CRN) – A network obtained from multiple CRs, which adjusts the network behaviour in response to the changing environment of the network in the event of a disaster.

7.2 Introduction

A Cyber Physical System (CPS) is a term coined to represent an amalgamation of network, computation and physical processes. It is a process of communication amongst the physical and computational constituents through networks. In cyber physical systems, embedded computers and networks handle and manage the physical processes with a feedback loop, whereas processes influence computations and *vice versa*. In general, CPS combines the working of the physical processes with networking and software, thereby offering generalization, design and modelling, and it develops novel techniques for the combined system.

Today, CPS plays an important role in areas like healthcare, emergency response, traffic flow supervision and electric power production and delivery, and the like. In the present work, a disaster response network is depicted as the CPS as the system can acclimate to continuously changing conditions during a disaster situation. Figure 7.1 depicts the concept map of a disaster response system. It shows the way in which the disaster response system is considered as a CPS. The disaster response system is recognized by the implementation of CPS to underlying networks in order to provide constant services to users based on their requests in disaster situations. During

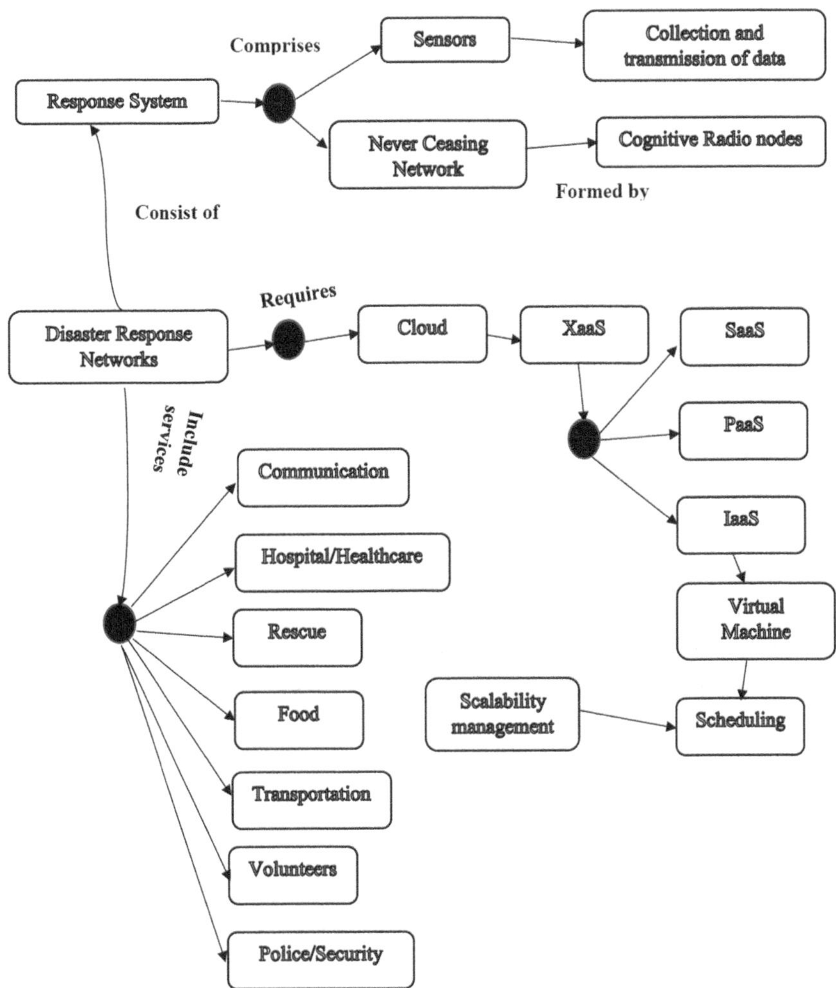

FIGURE 7.1
Concept map of the present work

large-scale disasters, offering services to victims in the disaster site is particularly important. The inspiration for our research originates from such experiences of natural calamities. In previous years, there has been a considerable rise in the number of natural calamities such as floods, earthquakes, forest fires and tsunamis across the globe. Though India is providing necessary measures to prevent loss of life by means of pre-disaster alert/warning systems, the entire communication network could have been severely damaged under such circumstances. A complete disconnect of the network leads to severe problems for evacuation and rescue activities, such as transportation of food, medicines and rescue materials. Disasters are spontaneous events which cause severe damage and/or loss of life. It would be essential to swiftly set up a new infrastructure which offers immediate services by using existing network resources, working in a heterogeneous environment. Numerous standardization activities and research practices have been carried out recently to manage disaster-resilient communication.

In recent times, there has been increasing interest in Internet of Things (IoT) systems, in which numerous sensors/devices are connected with the Internet to afford continuous services, devoid of any human interruption Sensors play an important role in disaster management systems. During a disaster, sensors can trigger the Never-Ceasing Network (NCN), to enable communication under any disaster situation. NCN is a communication network, which is rapidly installed as a response to a disaster, providing essential services to victims of the disaster. The Never-Ceasing Network plays a vital role in disaster management systems by providing constant connections to the services, offered through the cloud.

NCN must have the following characteristics:

- Should be a constant source of communication during a disaster.
- Offers quick data access.
- Has ability to adapt itself to network changes, due to congestion.
- Has ability to reorganize the network, if the network is disturbed.
- Ability to extend the network to attain greater network coverage.
- Ability to operate during disasters and normal times.
- Has independent power supply.

The NCN is a cognitive radio network which observes the operating environment, formulates decisions, configures the system based on observations and learns from experience obtained. A cognitive network has the ability to adapt to network changes, according to the observations. A cognitive network provides support for self-organization, interoperability, heterogeneity and re-configurability with available networks. This chapter focuses on the development of a prototype of the Never Ceasing Network in disaster-response networks.

7.3 Research Background and Objectives

7.3.1 Cyber Physical Systems

An amalgamation of network, computation and physical processes was depicted as a cyber physical system. It is a process of communication amongst the physical and computational constituents, through networks. After a natural calamity, there is a need for constant communication, with increases and changes in nature, owing to variable environmental constraints and requests from users. Thus, to offer information communication services during a disaster, information communication with intelligence was amended to observe the constantly changing situations. Intelligence was incorporated into the ICT system, consequently, so that the network was able to adapt to environmental changes and to changes in user demand. (Nishiyama, Suto, and Kuribayashi 2017) considered network of Access Points as physical processes. Machine-to-Machine communication was offered by implementing a Data Acquisition System (DAQ). The optimal gateway selection algorithm was implemented as a cyber part by using the user traffic information at every access point. The process of altering the gateway in the network was implemented as a control point. The optimized network capability under the variable conditions during the disaster was obtained, with the implementation of CPS to networks without the utilization of extra communication resources. Nowadays, natural disasters are considered to be an important research subject for the development of networks. In order to detect landslides, a data-driven CPS was adopted (Liu et al. 2018) developed the system for the detection of landslide and saved energy usage by decreasing data transmission to an extent. Wi-Sun acceleration sensors were used to detect the acceleration of the nearby environment in a 3D domain and the sensors were interconnected to a router. The correlation between the data samples obtained from each sensor was investigated and a data-driven traffic deduction mechanism was proposed by exploring correlations between the data from different sensors, which diminished the data volume considerably and assured the detection performance.

7.3.2 Need for a Disaster Response System and Scheduling

Disasters cause interruption to the information and communication systems which, in turn, make the system inaccessible when there is a peak demand for communication. At the time of the disaster, the people affected and victims are unable to communicate, either with their relatives or the rescue team. The disaster management systems should be designed in such a way as to provide open and prompt communication and information sharing between the various services, like first aid, food, hospital services and fire services. A distributed cooperative IoT system for flood disaster prevention employed an event-driven data collection system to effectively collect data from the Never-Die Network (NDN) to demonstrate the disaster resilience of the network (Kitagami et al. 2016). In addition, the rule-based autonomous control mechanism issues a local alarm for flood even when the IoT gateway is connected with the IoT cloud server. To evaluate the effectiveness of the proposed system, the experiments, were conducted using the original data of the river in Japan, and performed a field trial in Quang Nam Province, Vietnam. The distributed cooperative IoT system for flood disaster prevention reduced the network load for flood monitoring and provided flood warning at the appropriate time. The process of following the disaster-prone areas, which were constantly changing, was considered to be a difficult task. Thus, the system should be designed to offer adequate network capability as a service under a both normal and disaster conditions. The Never-Die Network was selected as the novel network system to provide the aforesaid service to the end-users. There was a need to monitor the network performance parameters, such as packet loss rate and throughput, in the disaster-affected areas (Sato et al. 2016). A method was proposed to monitor the packet flow by considering the throughput and packet loss in a system, whilst monitoring the state of multiple distinct access networks.

NDN is considered to be a robust network as it was able to tolerate sudden degradations or flux in network quality. The proposed NDN comprised a Cognitive Wireless Network and a Satellite Network. From the different wireless links, the researchers selected the best links and routes by employing the Analytic Hierarchy Process (AHP) and the Ad hoc on Demand Distance Vector (ADDV) method, respectively (Uchida and Takahata 2012). In order to minimize the disaster loss incurred, precise prediction and quick response were required. The Peer-to-Peer (P2P) cloud network services were used for the IoT-based disaster information systems (Kishorbhai and Vasantbhai 2017). The P2P network has the ability to upscale the increasing system capability during disaster situations. the present network was interconnected with IoT/Machine-to-Machine (M2M) network and provided the required services *via* the P2P cloud network. The network functioning of infrastructure layer was enhanced as a virtual network layer and a physical network. With the intention of managing disaster information systems, three plans, namely the peer distribution plan, the information integration plan and the load-balancing plan in the middleware layer were configured.

Today, IoT technology has the ability to be applicable during disaster situations (Sato et al. 2018). An IoT-based solution was developed to satisfy the user needs during disaster operations. The solution obtained was validated with a task-technology fit approach, which examined the importance of adopting IoT technology for disaster management. During disaster conditions, communicating networks were diminished and the restoration of networks was a puzzling issue. IoT offers communication in everyplace, everywhere and anytime. Under these circumstances, IoT offered information about resources which could be scheduled for distinct tasks. The resource scheduling problem was analyzed, using the banker's algorithm, and attained the optimum usage of resources (Kumar et al. 2017). In addition, the results obtained were assessed in terms of execution time and fairness of resource allocation. A cloud scheduling algorithm was developed, which took account of the existing Virtual Machine (VM) resource utilization by examining prior VM utilization levels. The results obtained were compared with traditional schedulers, available in OpenStack. The results showed that performance degradation decreased by 19% and CPU utilization increased by 2% when the cloud scheduling algorithm was used (Sotiriadis et al. 2018).

In the present work, the disaster management system is a combination of conventional network standards, with cloud computing and IoT technology being designed and built. In addition, scheduling of accessible resources for different user requests during a disaster has been recognized as a significant issue. In order to solve these issues, the following strategies were followed:

- A resilient network with suitable WSN architecture for collecting data from sensors was developed.
- A complete network capability to users during disaster, without any human interruption, was offered.
- Continuous, accessible cloud services, based on the user requirements, were available.

7.4 Purpose

The intention of the present work is to provide continuous services to users, according to their requests, through the Never-Ceasing Network system. NCN is built using a Cognitive Radio Network to offer communication under any disaster situations. The NCN is a novel network paradigm to provide network competence under both regular and disaster-based conditions. The following are the problems faced by information and communication systems during a disaster:

- Failure of the network in the existing communication and network infrastructures, such as wired and mobile communications.
- Inability to connect and transmit information, data, resources and services to the victims.
- Complete failure of network administrators.
- Since the network is dead, there is no possibility of serving evacuated regions of the disaster area.
- Power supply equipment failure and low battery power.

In order to resolve the aforesaid concerns, the network system should possess the following characteristics:

- Tangibly redundant hardware arrangement.
- Ability to recognize the changes in the network and reconfigure them automatically.
- Ability to expand the network coverage, to satisfy the user needs.
- Ability to provide network connection under both normal and disaster conditions.

- Independent power supply.

Instead of depending on a particular information communication unit, a network should be able to use a range of diverse information communication units. A different range of information and communication offers an integrated network management to easily follow the NDN situation in the case of disaster. The environment around the network varies from time to time during a disaster. Consequently, the wireless network systems need to be reconfigured to adjust, according to the changes in the network. NCN was shaped by means of the cognitive wireless network. A Cognitive Radio (CR) observes its environment, makes decisions and reconfigures, according to the observations, and learns from the experience. A Cognitive Radio Network (CRN) is obtained from multiple CRs and adjusts the network behaviour in response to the changing environment of the network in the course of a disaster.

7.5 System Architecture

Water level, temperature and humidity sensors were deployed to measure the flow speed and water level. Star topology is utilized for communication among sensors. The collected data from the sensors are sent to an IoT gateway *via* a sink node. The sensor detects the environmental changes periodically, and subsequently transmits the data to the sink node, using constrained application protocol (CoAP) messages at specific time intervals. The sink nodes are allowed to transmit the data to the IoT gateway, according to the available backhaul connectivity. The IoT gateway accepts diverse datasets from several sensor nodes and transforms them to a regular format. In addition, it performs protocol transformation by using various protocols for outbound communication that connect the gateway to the destined network. Once the threshold level is reached, the gateway sets off the Never Ceasing Network to operate. Once the NCN is live, the users are allowed to connect, to use the cloud as needed. In order to attain the disaster leniency of the network, the Never Ceasing Network was developed. Also, the Cognitive Radio Network was used to construct the resilient network during the disaster, without any human disruption. Services from different providers are registered in a service registry. Virtual Machine (VM) allocation includes the process of fair scheduling of the VM to satisfy the bulk requests from users during the disaster. Figure 7.2 shows the conceptual system architecture.

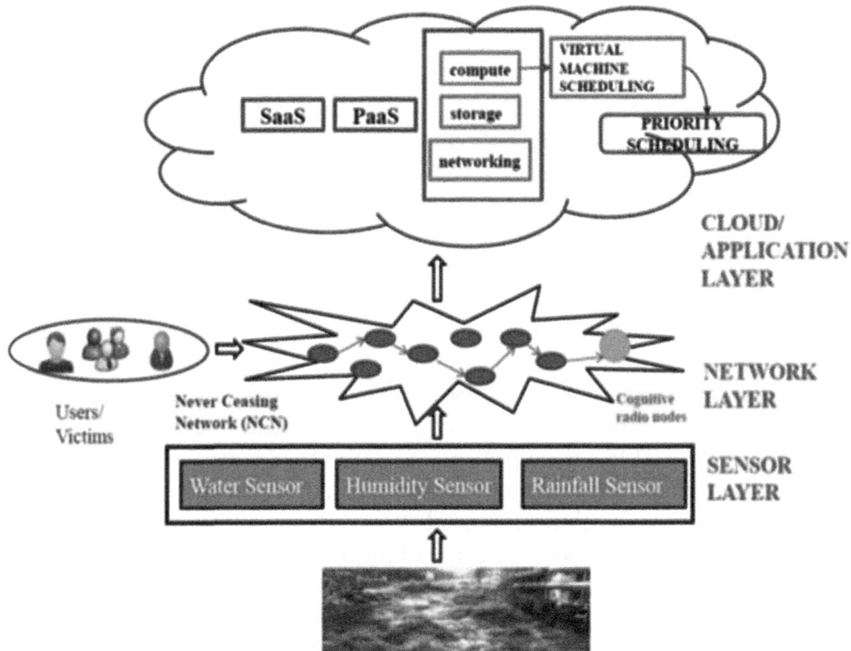

FIGURE 7.2
Conceptual system architecture

7.6 System Prototype Development

In this segment, the anticipated system prototype for the development of NCN is discussed. The proposed system uses IEEE 802.15.4 as the physical and datalink layer, IPV6 over low-power wireless personal area networks (6LoWPAN) as the network layer, User Datagram Protocol (UDP) as the transport layer and Constrained Application Protocol (CoAP) as an application layer for communication among sensor and sink nodes. The main issue of transferring the data from the sensor nodes to the IoT gateway is the inaccessibility of Wi-Fi/LAN connectivity in distant places. As a consequence, Wi-Fi/LAN is considered as backhaul connectivity in this system. The present work employs open-source platforms and consistent protocols to achieve diverse wireless sensor network systems.

7.6.1 Tools

Four components, namely Arduino Mega, XBee Shield, XBee Series 1 module and DHT22 sensor, are used to develop the Wireless Sensor Networks.

To measure temperature and humidity, the DHT 22 sensor is used. It uses a capacitive humidity sensor and thermistor to determine the surrounding conditions and outputs a digital signal. DHT 22 is a less-expensive sensor operating at 3–5V DC voltage with 2.5mA maximum current usage. Temperature measurement range is −40 to 80°C (±0.5°C). Humidity measurement range is 0–100% with 2–5% accuracy. The DHT22 sensor has four pins. The first pin on the leftmost is connected to 3–5V, the second pin to the data input pin and the rightmost pin to the earth of Arduino Mega. Arduino Mega is a microcontroller, based on ATmega1280, with 54 input/output pins. Of these, 14 pins are used as PWM pins, with 16 analogue pins, and four UART pins. It consists of a 128KB flash memory. Of the 128KB memory, 4KB is used by the boot loader, with an SRAM of 8KB and an EEPROM of 4KB. Arduino Mega is powered either by USB connection or with an external power supply. XBee Series 1 is a famous 2.4GHz XBee module from Digi, with a 250kbps data rate. It uses 802.15.4 to communicate with other nodes. XBee Shield provides a wireless communication between the Arduino board and the XBee Series 1 RF module. It allows the modules to communicate to up to 100 feet indoors or 300 feet outdoors. Arduino Ethernet Shield 2 is an open-source tool which offers communication between Arduino and the Internet. Three components, namely Raspberry Pi 2, XBee Series 2, and Wi-Fi Dongle are used to develop the IoT gateway. Raspberry Pi 2 consists of a quad-core Cortex A7 CPU, running at 900MHz with 1GB RAM. Raspberry Pi 2 includes Broadcom Dual-Core Video Core IV Media Co-Processor at 250MHz, with 40 pins extended to GPIO. XBee modules are reliable and offer simple communication between microcontrollers and computers. The development of NCN with cognitive radio nodes is shown with NetSim (Network Simulation). NetSim is an emulation tool which permits interfacing with devices, such as Raspberry Pi and Arduino. On the whole, a reliable, vendor-non-specific and cost-efficient heterogeneous WSN is obtained. The WSN consumes backhaul connectivity to transmit data from the sensor to the IoT gateway. Cloudsim is an open-source simulation tool for simulating cloud applications and associated algorithms. Figure 7.3 (a) and (b) show the sensor node components and the IoT gateway components.

7.7 Disaster Response Networks

Generally, disaster management is categorized into four stages, namely mitigation, preparation, response and recovery. The mitigation and preparation phases are organized prior to the occurrence of the disaster. In the course of the mitigation phase, information drives were presented, regarding public awareness, education, guarding and enhancing the prevailing infrastructure. For the period of the preparation phase, civic

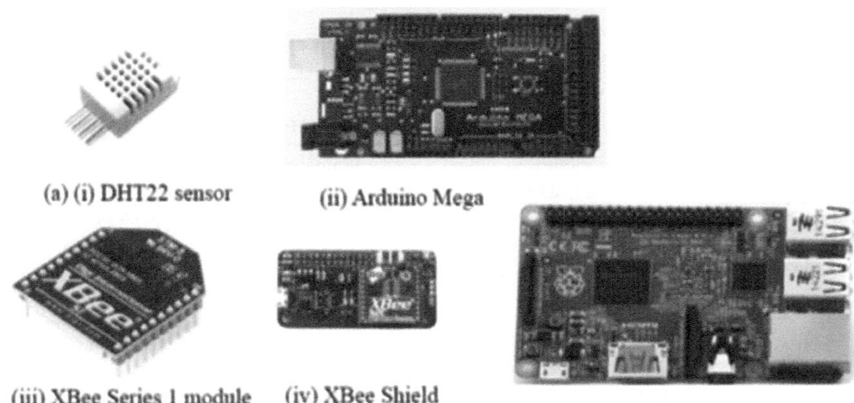

FIGURE 7.3
(a) Sensor node components: (i) DHT22 sensor, (ii) Arduino Mega, (iii) XBee Series 1 module and (iv) XBee Series 1; (b) IoT Gateway components: (i) Raspberry Pi, (ii) XBee Series.

alertness, volunteer supervision and disaster response strategy were set up. Immediately after the disaster struck, the response phase includes saving people, offering them basic needs (food, clothes, shelter, healthcare and security), with resource scheduling and resource provisioning being deployed. The recovery phase involves the recovery of all aspects of the disaster's impact on society and restoration of the normal economy (Uma et al., 2014). In post-disaster management, the scheduling phase is said to be the most important process. To handle the disaster situation in a real-time response, the different tasks need different resources (Uma et al., 2014). Consequently, resource availability needs to be scheduled in a specific order to accomplish the tasks within a definite time frame. During the disaster, the dynamic requests and frequently variable environment needs the involvement of a new technology to make an efficient and appropriate decision in minimum time (Subramanian et al., 2012). At the present time, cloud computing has become the most comprehensively used systems for providing virtual resources to the end users. Cloud computing delivers "Everything as a Service (XaaS)" (Nzanywayingoma and Yang 2018). The most vibrant process of such a system was termed task-based resource scheduling by using the Precedence Scheduling algorithm (PSA). During the disaster, the following tasks were considered for starting the network: providing medical facilities for the victims, evacuation rescue, restoration processes and the like. With the aim of achieving the aforementioned tasks, resources, like food, volunteers, fire services, hospital facilities and rescue teams, were required. The resources were provisioned in accordance with precedence, which was calculated based on the level of user requests and association with a specific resource.

TABLE 7.1

Scheduling Algorithms

Scheduling Algorithm	Explanation	Merits	Demerits
Round Robin (Prajapati and Raval 2013)	Pre-emptive algorithm, which allocates the tasks on the available VMs in a cyclic manner, with the task being allowed to store in a ring queue. Each task was allocated a quantum of time and, if it cannot be completed within its time, then it will be pre-empted and stored back at the tail of the queue and allowed to wait for its next turn. This algorithm repeats until each task in the queue was assigned with the needed resource.	Simple rule: allocate resources equally among the tasks. Emphasis on equity among the scheduled tasks; tasks are executed in turn and must wait for the previous task to finish before execution (starvation free).	Long tasks might take a longer time for complete execution, Servers were overloaded and pre-emptive policies relied heavily on the length of time slice, and, thus, the short time slice may lead to much switching.
First Come First Serve (FCFS) (Liu and Qiu 2016)	Queue follows the FIFO mechanism (first in, first out). The incoming tasks reach the queue with the shortest waiting time.	Simple. Easy to understand.	Non pre-emptive. Follows a single criterion for scheduling and the short jobs at the back of a queue might wait until the long tasks in the front of queue are completed.
Shortest job first (SJF) (Salot 2013)	Sort the set of tasks by placing the smaller tasks at the front of the queue and the longer tasks at the end of the queue.	Reduce average waiting time.	Starvation for longer jobs.

7.8 Scheduling Algorithms

In this section, the task-based resource scheduling and provisioning algorithms will be discussed, based on factors such as time, cost, power, resource usage, scalability and reliability, throughput, response time, migration time and security (Table 7.1).

7.8.1 Precedence Scheduling Algorithm

Algorithm: Pseudo code for Precedence Scheduling algorithm

Require: User submission of jobs

Ensure: Scheduling of jobs, based on resource requirement

1. Calculating the vector of criterion weights.
   1.1 Creation of pairwise comparison matrix M, n×n real matrix, where n is the number of evaluation criteria considered
   1.2 The entry m_{jk} of the matrix M depicts the importance of the j-th criterion relative to the k-th criterion
   1.3 The entries m_{jk} and m_{kj} satisfy the following constraint: $m_{jk} \cdot m_{kj}=1$

   criterion weight vector $w = \sum_{l=1}^{n} \frac{a_{jl}}{n}$
2. Calculating the matrix of option scores.
   2.1 The matrix of option scores is a m×n real matrix B. Every record b_{ij} of B denotes the score of the i-th option with regard to the j-th criterion.
   2.2 The score matrix B is obtained as B= $[b^{(1)} \ldots b^{(n)}]$, i.e., the j-th column of S corresponds to $b^{(j)}$
3. Positioning the options.
   3.1 When the weight vector w and the score matrix B have been computed, the vector P of global scores was obtained by multiplying S and w,
   3.2 $P = S \cdot w$ (P precedence vector)

Prior to scheduling of any task, the scheduler needs to collect the request, consisting of several tasks. It is important to obtain information concerning the physical resources, for example, the number of cores and the amount of available memory to allocate to all tasks consistently. At first, the jobs were sorted in a certain order, based on the precedence value (P) acquired from the Precedence Scheduling algorithm. The job with maximum precedence was offered highest importance and was scheduled first.

Steps for finding precedence

Input: Resource matrix

i. Resource matrix:

1	5	7
0.2	1	3
0.14	0.33	1

ii. Steps to calculate precedence:

a. Add each column of the matrix to acquire

1	5	7
0.2	1	3
0.14	0.33	1
1.34	**6.33**	**11**

b. Divide every constituent of the matrix by the sum of its column.

0.74	0.78	0.63
0.149	0.15	0.27
0.10	0.05	0.09

c. Average across the rows to find normalized principal eigenvector or

Precedence vectors

$$\text{Precedence Vectors} = \begin{matrix} 0.731 \\ 0.189 \\ 0.080 \end{matrix}$$

d. To check the consistency, eigenvalue (λ_{max}) is calculated

$$\lambda_{max} = (0.731)(1.34) + (0.189)(6.33) + (0.080)(11)$$
$$= 3.064$$

e. To find the Consistency Ratio (CR):

$$CR = CI/RI$$

where,

CI = Consistency index

RI = Random Consistency index

$$CI = \frac{\lambda_{max} - n}{n-1} = \frac{3.064 - 3}{3-1} = 0.03$$

The choice of RI is a span of the sequence of (0.00, 0.00, 0.58, 0.09, 1.12, 1.24, 1.32, 1.41, 1.45, and 1.49) and RI values for different numbers of n from 1 to 10.

$$CR = 0.03 / 0.58 = 0.05$$

f. If CR $\leq$ 0.10 (10%), then the measure of consistency is acceptable.

Here, 0.05 $\leq$ 0.10. Thus, the measure of consistency is acceptable.

The Precedence Vector obtained for the resource matrix is shown as follows:

Resource	Compute	Storage	Network	Precedence
Compute	1	5	7	0.731
Storage	0.2	1	3	0.189
Network	0.14	0.33	1	0.080

Steps to find the appropriate service with respect to available resource

Input: Job matrix: The job matrix is two-levelled with jobs and respective resources.

1. Job matrix:

$$\begin{matrix} 1 & 7 & 5 \\ 0.14 & 1 & 1 \\ 0.2 & 1 & 1 \end{matrix}$$

a. Average across the rows to find normalized principal eigenvector or

Precedence vectors

$$\text{Precedence Vectors} = \begin{matrix} 0.74 \\ 0.12 \\ 0.14 \end{matrix}$$

b. Eigenvalue (λ_{max})=3.012

c. CR=0.013<=0.1

Thus, the measure of consistency is acceptable

2. To compute precedence vectors for available services, namely Rescue, Hospital, Food and Volunteer, with regard to available resources, namely compute, storage and network correspondingly.
3. Precedence vectors for Rescue, Hospital, Food and Volunteer services relating to computed resource

$$\text{Precedence Vectors} = \begin{matrix} 0.64 \\ 0.21 \\ 0.08 \\ \\ 0.072 \end{matrix}$$

d. Eigenvalue (λ_{max})=4.15

e. CR=0.05<=0.1

4. Precedence vectors for Rescue, Hospital, Food and Volunteer services, relating to storage resource

$$\text{Precedence Vectors} = \begin{matrix} 0.37 \\ 0.43 \\ 0.13 \\ 0.07 \end{matrix}$$

f. Eigenvalue (λ_{max})=4.04

g. CR=0.017<=0.1

5. Precedence vectors for Rescue, Hospital, Food and Volunteer services, relating to network resource

$$\text{Precedence Vectors} = \begin{matrix} 0.63 \\ 0.23 \\ 0.06 \\ 0.08 \end{matrix}$$

h. Eigenvalue (λ_{max})=4.20

i. CR=0.07<=0.1

6. Ranking of alternatives:

$$\begin{matrix} 0.64 & 0.37 & 0.63 \\ 0.21 & 0.43 & 0.23 \\ 0.08 & 0.13 & 0.06 \\ 0.072 & 0.07 & 0.08 \end{matrix} \quad * \quad \begin{matrix} 0.74 \\ 0.12 \\ 0.14 \end{matrix}$$

↓ ↓

Criteria weights

$$= \begin{matrix} 0.6062 \\ 0.2392 \\ 0.0832 \\ 0.072 \end{matrix}$$

Here, Rescue service is handled with the resources requested, as it ranks first.

TABLE 7.2

Comparison of Various Scheduling Algorithms in a Cloud Environment in Accordance with the Following Constraints

S.NO	Algorithm	Throughput	Response Time	Migration Time	Time Efficient	Resource Usage	Scalability and Reliability	Power Aware	Cost-Effective	Security	Memory-Aware/ Bandwidth-Aware
1.	Round Robin	✓	✓		✓	✓	✓				
2.	FCFS					✓	✓				
3.	Match-making algorithm	✓				✓	✓		✓		
4.	Shortest job scheduling	✓	✓		✓	✓	✓				
5.	Precedence Scheduling algorithm	✓	✓		✓		✓	✓	✓	✓	

7.9 Comparative Analysis

The practice of choosing the correct algorithm for the scheduling of tasks to resources depends heavily on aspects like time, power, throughput, response time, security, resource usage and the like. Table 7.2 shows the comparison of various scheduling algorithms in the cloud environment, in accordance with the constraints shown (Table 7.2).

7.10 Conclusion

Scheduling and Resource provisioning were considered to be the most important factors during disaster situations. The implementation of the cyber physical system includes the working of IoT and the cloud environment for effective communication between the entities, which helps achieve efficient utilization of the available resources. The proposed algorithm schedules the tasks in view of the precedence calculated, based on user interest and demand for a specific resource. The proposed approach was estimated in terms of time, power, throughput, response time, security, resource usage and the like, which exhibited better results than the approaches stated in the literature. The proposed algorithm can be further extended by future work, with the implementation of machine learning algorithms which, in turn, could positively influence the performance of the overall system.

References

Kishorbhai, Vasani Yash, and Nagekar Nainesh Vasantbhai. 2017. "AON: A Survey on Emergency Communication Systems during a Catastrophic Disaster." *Procedia Computer Science* 115: 838–845. doi:10.1016/j.procs.2017.09.166.

Kitagami, Shiji, Vu Truong Thanh, Dang Hoai Bac, Yoshiyori Urano, Yohtaro Miyanishi, and Norio Shiratori. 2016. "Proposal of a Distributed Cooperative IoT System for Flood Disaster Prevention and Its Field Trial Evaluation." *International Journal of Internet of Things* 5(1): 9–16. doi:10.5923/j.ijit.20160501.02.

Kumar, J. Sathish, Mukesh A. Zaveri, and Meghavi Choksi. 2017. "Task Based Resource Scheduling in IoT Environment for Disaster Management." *Procedia Computer Science* 115: 846–852. doi:10.1016/j.procs.2017.09.167.

Liu, Li, and Zhe Qiu. 2016. "A Survey on Resource Scheduling in Cloud Computing." *2016 2nd IEEE International Conference on Computer and Communications* 13(9): 2717–2721. doi:10.1007/s10723-015-9359-2.

Liu, Zhi, Toshitaka Tsuda, Hiroshi Watanabe, Satoko Ryuo, and Nagateru Iwasawa. 2018. "Data Driven Cyber-Physical System for Landslide Detection." *Mobile Networks and Applications*, 1–12. doi:10.1007/s11036-018-1031-1.

Maheswari, J Uma, and G R Karpagam. 2014. "Ontology Based Comprehensive Architecture for Service Discovery in Emergency Cloud." *International Journal of Engineering and Technology* 6(1): 242–51.

Maheswari, J Uma, G R Karpagam, and A Bharathi. 2014. "Self-Organizing Agent Based Framework for Service Discovery in Cloud."

Nishiyama, Hiroki, Katsuya Suto, and Hideki Kuribayashi. 2017. "Cyber Physical Systems for Intelligent Disaster Response Networks: Conceptual Proposal and Field Experiment." *IEEE Network* 31(4): 120–128. doi:10.1109/MNET.2017.1600222.

Nzanywayingoma, Frederic, and Yang Yang. 2018. "Efficient Resource Management Techniques in Cloud Computing Environment : A Review and Discussion Review and Discussion." *International Journal of Computers and Applications* 7074(January): 1–18. doi:10.1080/1206212X.2017.1416558.

Prajapati, KD, and P Raval. 2013. "Comparison of Virtual Machine Scheduling Algorithms in Cloud Computing." *Journal of Computer* 83(15): 12–14. doi:10.5120/14523-2914.

Salot, Pinal. 2013. "A Survey of Various Scheduling Algorithm in Cloud Computing Environment." 131–135.

Sato, Goshi, Noriki Uchida, and Norio Shiratori. 2018. "Complex, Intelligent, and Software Intensive Systems." 611. doi:10.1007/978-3-319-61566-0.

Sato, Goshi, Noriki Uchida, Norio Shiratori, and Yoshitaka Shibata. 2016. "Implementation and Evaluation of Never Die Network System for Disaster Prevention Based on OpenFlow and Cognitive Wireless Technology." In: *Proceedings 2016 10th International Conference on Complex, Intelligent, and Software Intensive Systems, CISIS 2016,* 190–196. doi:10.1109/CISIS.2016.113.

Sotiriadis, Stelios, Nik Bessis, and Rajkumar Buyya. 2018. "Self Managed Virtual Machine Scheduling in Cloud Systems." *Information Sciences* 433–434: 1339–1351. doi:10.1016/j.ins.2017.07.006.

Subramanian, S, G Nitish Krishna, M Kiran Kumar, P Sreesh and G R Karpagam. 2012. "An Adaptive Algorithm for Dynamic Priority Based Virtual Machine Scheduling in Cloud." *International Journal of Computer Science Issues* 9(6): 397–402. https://www.ijcsi.org/papers/IJCSI-9-6-2-397-402.pdf.

Uchida, Noriki, and Kazuo Takahata. 2012. "Never Die Network Based on Cognitive Wireless Network and Satellite System for Large Scale Disaster." *Journal of Wireless Mobile Network Ubiquitous Computing and Dependable Applications* 3(3): 74–93. http://isyou.info/jowua/papers/jowua-v3n3-5.pdf.

8

Risk-Adjusted Digital Supply Chains

Sumit Kumar and Amit Kumar

CONTENTS

Organization of the Chapter

Section 1 presents the terms and terminologies for the user to understand the chapter. **Section 2** introduces the concept of supply chain execution. In **Section 3**, we will talk about the risks and variabilities commonly encountered in supply chains. **Section 4** explains how risk impact prediction can be conducted through dynamic simulation. **Section 5** suggests leveraging simulation and optimization for the next-best action recommendation. The challenges and opportunities in using one such system are explained in **Section 6**.

8.1 Terms and Terminologies

- Internet of Things (IoT): A system of interrelated computing devices, mechanical and digital machines, objects, animals or people that are provided with unique identifiers and the ability to transfer data over a network without requiring human-to-human or human-to-computer interaction.

- Key Performance Indicators (KPI): A set of equations/calculations based on the process outcome used to measure the performance of a system.
- Dynamic simulation: A method whereby the current state of the system or "snapshot" is constantly read and used as an input for simulation through time.
- AnyLogistix: A simulation platform built specifically for large and complex supply chains. AnyLogistix is a product of the AnyLogic Company.
- Enterprise Resource Planning (ERP): Refers to the integrated management of a company's business processes, using software and technology.
- Stock-Keeping Unit (SKU): The identification of a product for inventory management and traceability purpose. It is usually an alphanumeric code assigned to the product.
- Out-Of-Stock (OOS) or stock-out: Refers to a situation where a product is demanded by the customer but is not available or the inventory level is zero.
- Business Intelligence (BI): In this chapter, BI refers to certain visualization tools that take input data from Excel or ERP and make it possible to create a dashboard that shows the key KPIs of the supply chain.

8.2 Introduction

Due to the fierce competition in the market, supply chains are becoming increasingly complex. Latest technologies are continually being evaluated and used by companies to bring value to the end-customer. Efforts to convert factories to smart factories or to achieve digitization of factory and supply chains are currently underway. With the fourth industrial revolution underway, the focus is on using Internet of Things (IoT), cloud computing and analytics to take the supply chains to the next level. One such application involves creating a digital twin of the supply chain, which is a Cyber Physical System (CPS) for the supply chain and which can be used to predict the near-future state of the supply chain (Iris Heckmann & Stefan Nickel 2017 and Iris Heckmann 2016).

Once we are able to predict the future state, then any disruptions or risks can be anticipated, and corrective actions can be taken. It is also possible to make the digital twin model suggest the best course of action, by using prescriptive analytics. There are two requirements for creating a digital twin of the supply chain. The first is to have a simulation model of the entire supply chain at a sufficient level of detail so as to be useful for the purpose. The second requirement is for the simulation model and the physical supply chain to be connected by IoT. Once this cyber physical system is ready and updated with the latest information/state of the physical supply chain, also called a "snapshot", multiple replications of the simulation model can be run to project and determine how the proposed planning decisions would perform. The digital twin would also incorporate all the variations and uncertainties present in the supply chain and all the external risks, such as weather, pandemic, political unrest, etc.

8.2.1 Solution Design

The cyber physical system of the supply chain that we propose can be seen in Figure 8.1. The current state or snapshot of the system can be taken from the ERP system, which has the information of where and how much of the inventory is in the supply chain. The ERP system will also provide in-transit inventory information and the production plan at the manufacturing facilities. We can take the demand data for the next couple of weeks from a demand-planning module or, better still, from a demand-sensing solution that predicts the demand at a more granular level, such as at a Stock Keeping Unit (SKU) and Distribution Center (DC) level for each day. This demand information includes the forecast error, which is used in the simulation. Other uncertainties in the system, such as disruptions and demand surges, are also inputs to the simulation model. The core of the solution is the simulation model itself. This model replicates the actual supply chain configuration in terms of the manufacturing facilities, distribution centres and the customers or demand points. It has all the business rules configured and hence is able to replicate the operations of the supply chain.

The simulation model is run multiple times, each run being called a replication. Multiple runs, or replications, are needed to see the behaviour of

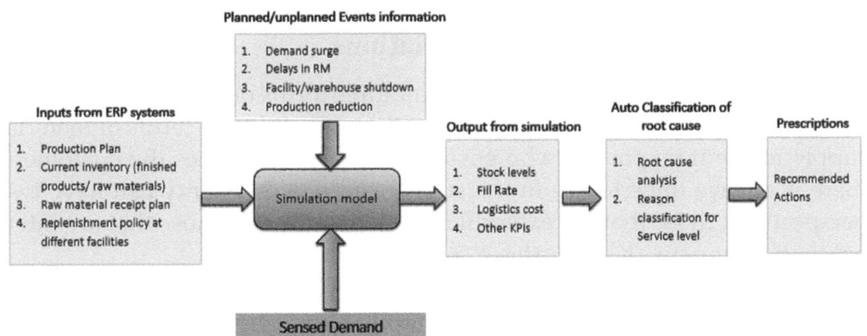

FIGURE 8.1
Overall solution design of the Cyber Physical System of the supply chain

the system under a situation of uncertainty. In each run, the model will run one scenario based on the input uncertainty and variation. The number of replications needed to arrive at a reasonable conclusion depends on the confidence level desired and the level of underlying uncertainty. The output of the simulation model are the key performance indicators (KPIs), such as out-of-stock (OOS), fill rate, cost, etc. The output will be in terms of the probability of stock-out. For example, there will be a 90% probability that SKU x will be stocked-out at distribution centre (DC) y in period z. This means that, in 90% of the runs, this stock-out occurred. Similarly, the logistics cost will be a range and can be plotted as a histogram or a box plot to see the logistics cost across all the replications (Borshchev A. 2013 and Ivanov D. 2018).

The output of the simulation model will be an input for an Artificial Intelligence (AI) system that will autoclassify the root cause of the stock-out or the low service level. This autoclassification algorithm is based on looking at the inventory patterns across time at the various stocking points and identifying the cause, based on a pretrained neural network (D Lowd & P Domingos 2005 and Satyendra Kumar Sharma & Saurabh Sharma 2015).

The final step of the solution is to recommend the next-best action, in order to mitigate the stock-out situation. This prescriptive action will have the objective of minimizing the total cost. For example, if a stock-out will happen with a high probability and if a nearby distribution centre has available inventory for that SKU, then the system might recommend inter-stock transfer between the distribution centres at an additional cost.

Implementing this solution in any supply chain will create a cyber physical system that will lead to a self-driving enterprise. We recommend the creation of such a system and to run it daily for a horizon of next two to three weeks. The system can also be run multiple times a day as and when any new event information such as lane closure, facility closure etc. become available.

8.3 Supply Chain Risks and Variations

A complex supply chain carries with it multiple forms of variations and risks, from the suppliers who are associated with unreliability in terms of material supply to the transporters who ensure the material reaches the customers, there is always uncertainty in the supply chain. Uncertainty can have two perspectives: those that are external to the company and those that are internal to the company (Kleindorfer, P. R. & G. H. Saad 2005).

Below are some of the typical failures in supply chains:

- Environmental risks, such as heavy rains, floods, other extreme weather conditions.
- Labour unavailability, due to strikes and slowdowns.
- Equipment breakdown at production or logistics.

- Distribution risks, such as logistics unavailability and warehouse closures.
- Political crisis or unrest.
- Supplier-related risks, or risks with contract manufacturers.

External events typically comprise disruptions, like inclement weather, political disruptions or even small-operation disruptions, such as fleet unavailability, supplier facility closure, etc. Whereas the major disruptions are rare and difficult to predict, the operational disruptions are predictable to a certain extent. More often than not, it is these short-term operational disruptions which keep impacting on our supply chain.

The other form of disruption is internal disruption, such as disruptions like operator unavailability and machine failures. These disruptions are either known or else can be predicted with reasonable accuracy from past data, using basic data analysis. A third, and probably the most chronic form of disruption, is the variation in a process. In a typical supply chain, variations can be seen in the form of demand variation, transportation lead time variation, processing time variation, etc. Variations are generally mitigated through buffers, such as safety stock, which is an inventory buffer, or safety lead time, which is a lead time buffer. However, all mitigation actions come with a severe cost on the operations and will significantly reduce the profit of the enterprise.

Some of the impacts apparent due to the operational risks are as follows:

- Excess logistics cost, as the standard options are not available due to disruptions.
- Stock-out at the distribution centre (DC), leading to lost sales.
- Inventory write-offs, due to incorrect placement of inventory.
- Sales opportunity loss.
- On-time delivery is affected, leading to lower customer satisfaction.

8.4 Predicting the Impact of Risks, Using Dynamic Simulation

A better, cost-effective and feasible approach to risk management can be achieved by predicting the impact of supply chain disruption through a cyber physical system. Most supply chains operate under a finite set of resources. This means that there is a finite number of warehouses, trucks, roads, factories and customers. Furthermore, most supply chains are governed by operation rules for inventory management, distribution, raw material sourcing, etc. What it does imply, then, is that a mirror or digital twin of the supply chain can be created and simulated to predict the outcome and impact possibilities within a short period of time. Since the supply chain operates within a finite boundary of rules and infrastructure, the possibilities, while variable, are still limited and can be calculated (Figure 8.2).

To this end, predicting the impact of disruptions and variations is very much possible through a robust simulation tool, which can mirror the entire supply chain and run simulations to visualize the impact. Since many of the events and data are probabilistic in nature, it makes sense to do multiple simulations to evaluate the range and consistency of the impact predictions (Figure 8.3).

Many of the KPIs and their performance can be evaluated through simulation. One of the prominent tools in this area is AnyLogistix, which can perform simulation of the entire supply chain network, considering each and every factor and the variances possible. Most of the KPIs commonly investigated in supply chains, related to revenue, inventory, transportation or service, can all be visualized on the tool and can also be readily exported into any BI tool for further visualization (Figure 8.4).

One important thing to remember is, how far into the future can predictions and simulations be made. We all know that, as we project further into the future, any forecast becomes more blurred and the error ranges increase. On similar lines, resource availability is also uncertain as we move into the future. The same also holds for many of the disruptions that can happen (Figure 8.5).

Hence, appropriate judgement has to be used while doing such simulations. The lead time for simulation cannot be so short that no actions can be taken, even with alerts, but can also not be so long that the alerts themselves

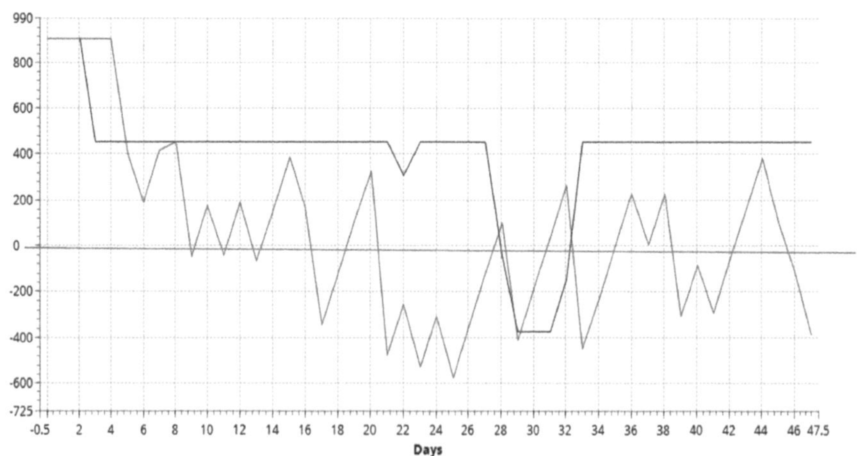

FIGURE 8.2
Probabilistic inventory profile for the future, achieved from simulation on AnyLogistix

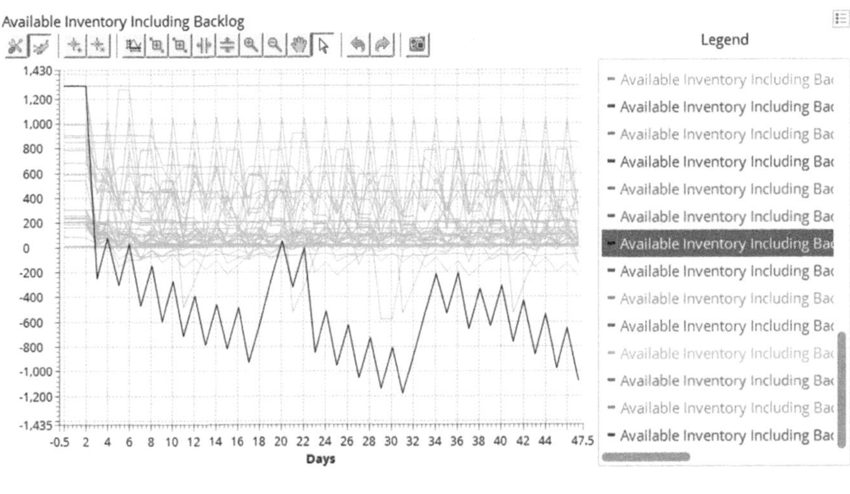

FIGURE 8.3
Projecting available inventory and backlog or lost sales, using AnyLogistix

#	Name	Event Type	Parameters	Occurrence Type	Occurrence Time	Trigger	Probability
1	Raining season	Path state	Path: Path Factory...	Random	7/1/19 - 8/30/19		0.5
2	End of raining sea...	Path state	Path: Path Factory...	Date	9/1/19 12:00 AM	Raining season	1
3	Increase in Demand	Demand coeffici...	Customer: [Custo...	Random	4/1/19 - 6/30/19		0.8
4	Decrease in Dema...	Demand coeffici...	Customer: [Custo...	Random	8/1/19 - 9/30/19		0.9
5	Factory failure	Facility state	Site: Factory Qui ...	Random	1/1/19 - 12/31/19		0.5
6	Factory recovery	Facility state	Site: Factory Qui ...	Delay (days)	7	Factory failure	1

FIGURE 8.4
Illustrative: capturing possible events and disruptions in the supply chain

have very low accuracy. The trade-off must be maintained, and this trade-off will be specific to each company and industry. Typically, running this simulation daily for the following two weeks can be appropriate for some supply chains.

8.5 Autoclassification of Root Cause and Proposal of Next-Best Action

Once a simulation has been carried out, the next step will be to take the appropriate actions to mitigate the impacts observed in the simulation

	Statistics name	Replication	Value	Unit
1	Service Level b...	Replication 10	0.4	Ratio
2	Service Level b...	Replication 1	0.52	Ratio
3	Service Level b...	Replication 2	0.49	Ratio
4	Service Level b...	Replication 3	0.53	Ratio
5	Service Level b...	Replication 4	0.55	Ratio
6	Service Level b...	Replication 5	0.5	Ratio
7	Service Level b...	Replication 6	0.59	Ratio
8	Service Level b...	Replication 7	0.58	Ratio
9	Service Level b...	Replication 8	0.54	Ratio
10	Service Level b...	Replication 9	0.55	Ratio

FIGURE 8.5
Service level estimation due to disruptions through multiple replications of the simulation

output. A typical supply chain would have many such supply chain performance failure alerts and it is important that we prioritize the alerts, based on the impact on the most important KPIs.

There can be two dimensions to failure alerts, namely the probability of failure and the impact of failure. A method needs to exist which can prioritize the alerts which need to be acted upon. However, in many cases, it may not even be possible to manually act upon each alert, considering the size and complexity of the network. In those cases, prediction of alerts alone will not serve the purpose and it becomes important to have a system supported by prescriptions for these alerts (Figure 8.6).

Once the alerts have been prioritized, the next step is to try and identify why the alert occurred or what was the root cause for the poor KPI. The main reasons for an OOS situation are the following:

- Production planning issues
    - Production plan issue
    - Production compliance or execution issue
- Distribution issues
    - Unbalanced distribution
    - Logistic issues, such as truck availability or load building
- Inventory issues
    - Forecast error, demand surge
    - Incorrect inventory norms

	Product				
Distribution Center	PPDV2	PPDV31	PPDV55	PPDS17	PPDS75
ML1201		100%		100%	100%
ML1287		97%			85%
ML1725		100%	100%		100%
ML2461		96%	40%		0%
ML2519		100%			20%
ML2593		52%			100%
ML2595	90%	100%	0%		100%
ML2641	0%	77%	90%		95%
ML2674		100%		100%	100%
ML2721		100%			100%

Expected service level of products at Distribution centers for next 2 weeks

FIGURE 8.6
Stock-out alerts provided for multiple days in the future

Based on the root cause, the problem can be highlighted to the person responsible, such as the production planner, the logistics planner or the demand or inventory planner. Once the root cause has been determined, the next step is to determine the next-best action possible to mitigate the OOS or to improve the KPI. Some of the typical proposals that can be provided by the system are:

- Advance shipment to selected locations, considering future disruptions.
- Movement of stock between warehouses to balance excesses and shortages.
- Alteration of production plan to prioritize shortages.
- Distribution plan modification.
- Sourcing alteration.

The applicability of each of these countermeasures will vary by company and by industry. What may be feasible for one situation might look impractical for another. Furthermore, suggesting a solution to the risk may still not be straightforward. In many cases, the overall failure pattern may be more important than the individual failure instance. Despite all these limitations in using preconfigured prescriptions, these could still be extremely helpful in many cases and take risk management to a practical stage beyond the theory.

8.6 Challenges and Opportunities

Risk management through cyber physical systems is a complex subject and there are challenges at every stage of the process. The challenges to carry out a precise risk assessment and mitigation come at different stages. Some of the most common challenges are:

- Identifying the risks: This is perhaps one of the key challenges to the entire risk management process and perhaps the most complicated part. The risks typically can be classified into various types and, with the support of advanced analytics and Web APIs, many of these can be addressed.
  - Internal risks: E.g. line failure, unavailability of personnel, etc. These risks are either known to the operating people or can be predicted with a reasonable probability by leveraging past history.
  - External risks: E.g. inclement weather, road closure or political disruptions. Here again, local knowledge comes in handy. At the same time, the use of predictions, which are available through paid domains, and connecting the data through web APIs will be useful.
- Risk mitigation: Here, the methods of addressing risks and their impacts are very contextual and, hence, standardization is very difficult. There are just too many variables under which a countermeasure can be decided.

However, despite all the hindrances, the benefits of risk management outweigh the efforts. Most companies lose significant business, revenue or margins, due to risks which are more operational in nature. Most of the time, these are missed and their impacts are just absorbed. A detailed analysis and solution development, leveraging tools supporting the concept of a cyber physical system, like AnyLogistix, and a careful risk management process setup, is important to realize the full benefits.

References

A. Borshchev (2013). *The Big Book of Simulation Modeling: Multimethod Modeling with Anylogic 6*. AnyLogic North America.

D. Lowd & P. Domingos (2005). Naive Bayes models for probability estimation. In: *Proceedings of the 22nd International Conference on Machine Learning*, 529–536.

Iris Heckmann & Stefan Nickel (2017). Rethinking supply chain risk analysis – Common flaws & main elements, Supply Chain Forum. *Anais an International Journal*, 18(2), 84–95.

Iris Heckmann. *Towards Supply Chain Risk Analytics: Fundamentals, Simulation, Optimization*. Springer Gabler Verlag.

D. Ivanov (2018). *Supply Chain Simulation and Optimization with any Logistix. 2nd, Updated Edition*. Berlin School of Economics and Law.

P. R. Kleindorfer & G. H. Saad (2005). Managing disruption risks in supply chains: *Production and Operations Management*, 14(1), 53–68.

Satyendra Kumar Sharma & Saurabh Sharma (2015). Developing a bayesian network model for supply chain risk assessment, *Supply chain forum: an International Journal*, 16(4), 50–72.

9

War Gaming for Cyber Physical Systems

Sujatha Nagarajan

CONTENTS

Organization of the Chapter

Section 9.1 presents the terms and terminologies for the user to understand the chapter. **Section 9.2** describes the origin of war gaming. **Section 9.3** depicts the relationship between video games and war gaming. **Section 9.4** shows the role of the cyber physical system in war gaming. **Section 9.5** summarizes the work and provides necessary observations.

9.1 Terms and Terminologies

- War Gaming: A methodology that aims to simulate the real-time Cyber Physical System in order to provide training to professionals and to empower them to understand and defend their systems in a highly interactive, immersive and fun-filled manner.

- Military War Gaming: A serious tool for training and research of generic and specific military tactical scenarios, where players command their troops towards victory.
- Educational War Gaming: Studying a domain by creating computer simulation games and learning from them.

9.2 The Origin

Historically, War Games are the most popular and established technique among military professionals during their training period, and are used to simulate battle techniques, play out complicated and multifarious attack and defence scenarios, and to meticulously plan out novel schemes for future battles. In the 19th century, the Prussians named this method, "Kriegsspiel" [1], a genre of War Gaming devised to teach military tactics to officers in command. The military War Games have mostly been of the physical model kind, with a paper map, miniature models and players, playing in teams or as individual players. War Games provided strategic decision-making battle experience before any major resource needed to be allocated for any upcoming war. This is due largely to the fact that games can efficiently reflect the core characteristics of war, such as conflict, competition, strategy, win and loss, in a painstakingly controlled and safe environment.

9.2.1 Military War Gaming

Military War Gaming is an age-old practice, which elaborately acclimates commanders to war-like simulations. In addition to that, it also assists them in the proactive preparation for future battles and wars, giving them an edge over their enemies. The games are generally designed in such a way that they have elaborate maps, physical miniature models and terrain to mimic the actual battlefields. Players, in this case military personnel, take the game seriously, as if it were a real war, and exercise the real-time chain of command, with vigorous competition, with the sole purpose being winning for their country. This gameplay paves the way for them to mentally process the strategies and to face the known and unknown conflicts, utilizing the information at hand. Subsequently, a post-mortem session accompanies the game, to conduct a discussion about the experience, decision making and strategies followed, and also to strengthen the existing simulation.

9.2.2 Educational War Gaming

Educational War Games have been a popular technique used by academics to teach students about domain knowledge, and to engage them in an interactive way of learning new information. For example, Philip A.G. Sabin has been using simulated War Games [2] for over a decade, to teach war-related studies.

9.2.3 Cyber Physical System–War Gaming

Inspired by the military War Games and the educational War Games, similar games can be designed to complement the cyber security-strengthening pipeline. Cyber Physical System–War Gaming is based on the ideology of the Military War Gaming method and of Video Games. The well-established War Gaming experience can be brought into the Information Technology and Cyber Physical System infrastructure, where War Gaming can prove to be a supreme asset for the multi-layered cyber defence pipeline, including but not limited to vulnerability vector assessment, penetration tests and iterative tests.

9.3 Video Games and War Gaming

Games of any kind create a competitive environment, where players sharpen their skills, figure out new capabilities, engage well with other players and create a strong bond with the team. Though board games have evolved into graphics-focused video games, people still immensely enjoy the face-to-face interaction and the comradeship of fellow players. Online video games provide similar ambience, where players can interact with fellow gamers from any corner of the world. It has been widely debated whether video games cause addiction, aggression and depression. Though multifarious research has been conducted to test this hypothesis, it cannot be denied that video games support gamers, to help them deal with their personal difficult times. Setting that debate aside, video games can be shown to be an immensely efficient training tool, with great benefits.

9.3.1 Video Games for Training

Video Games, when observed through a non-gamer's lens, become an extremely powerful training medium. Games have the capacity to accurately represent the crucial real-time systems of any industry sector, e.g., warfare, biomedical sector, agriculture etc. The significance of games as training means has spread far and wide, and even brushed against Cyber Physical

Systems about a decade ago. Games are normally associated with frivolousness, due to the existing preconceived notion in many societies. The term "Cyber Physical System–War Gaming", when suggested as a serious training medium, might even humour some people, because games have been looked at as a playful toy that distracts individuals from real-world hard-core work. Less understood, though, is that, given the chance that it deserves, along with the skill set and resources it imparts, a perfected Cyber Physical System–War Game has the ability to provide an impactful addition to the cyber defence pipeline and its enrichment.

> The possibility is that games, in fact, are not distracting us from reality, but they are preparing us to tackle the world's most urgent challenges.
>
> – Jane McGonigal [3]

Far from being disruptive to the workplace zen, War Games have the gripping potential to add a highly interesting and interactive angle to cyber security training. Video games are fun, interesting and also highly immersive, with their accurately simulated gameplay, strong mechanics, excellent audio playback and proven high-range visual rendering. In the case of Military War Games and Educational War Games, they have predominantly been played as board games with resized models and detailed maps. It does not have to be the same with Cyber Physical System–War Games. The reason is that computer games have replaced board games for several decades, and their development is a highly streamlined process, using various game engines and tools. If there is a medium that teaches its players and facilitates them to gain knowledge and skills, without explicitly stating so, it is certainly video games. When treated casually, games end up being casual games, but on the other hand, when treated seriously by the personnel giving and taking the training, they can prove to be of tremendous value to the company involved, and a cost saver in the long run.

Upcoming generations need no introduction to video games and their popularity, especially in India. India is among the top five fastest-growing markets for mobile gaming [4], with male gamers playing about 10–20 minutes per gaming session, and female gamers about 8–12 minutes. With the aim of modelling a better cyber security architecture, video-games-inspired War Games would attract a younger workforce to actively participate, absorb and contribute to the security pipeline in the 21st century.

9.3.2 Gamification

Gamification is a term coined in 2002, by Nick Pelling, to denote the utilization of game design knowledge in creating game-like experience for business and education [5]. The word did not gain much traction until much later. In 2014, the word "gamification" was redefined by Gartner, Inc., the world's leading research and advisory company, as "The use of game mechanics and experience design to digitally engage and motivate people to achieve their goals" [6].

According to this definition and its explanation, Cyber Physical System–War Gaming can be achieved by the gamification of the cyber security pipeline.

Gamification is the future of technological development as well, as suggested by the Army War College, Pennsylvania, a 117-year-old institution, with their latest research breakthrough [7]. They are working on using Artificial Intelligence to assist humans, by playing within the military rules and regulations, and by training the combatants with AI-infused War Games.

9.3.3 The Need for War Games in Cyber Physical Systems

Users get the opportunity to engage in an interactive way and understand, in depth, the realism of the cyber security threats and the organizational structure followed during the process of mitigation. The Cyber Physical System–War Game can be designed in such a way that the players are most active and engaged, while also learning the technical nuances of the domain. Players should be able to see visible progress based on their actions and the supervisors should be able to track the game metrics, to determine the player's performance. Games should be designed to encourage and facilitate communication between inter-related department personnel.

The main question that would arise inside any rational mind would be: "Why do we need to consider an additional training module, which would require nontrivial resources, especially the most important one of all the resources, i.e., time?". The answer would be to keep up with the latest requirements in the cyber security domain, and to replace the raw-theoretical way of training, which can sometimes become boring, to be frank. A stimulated brain invents new ideas. The rapidly growing advances and research in the Information Technology and Cyber Physical Systems fields implicitly emphasize the inevitability of shielding the data, physical systems and their software. Why not chose a fun, exciting, impactful, and interesting path to do the same?

Information is power, which we are all aware of. Hackers target the networked user information and misuse them, mostly for monetary gain, which is also the bitter truth. In order to protect the systems from such criminal misuse, cyber security protocols are appropriately placed. These protocols are assessed and audited periodically, scaling at the company-wide level. The question is: are these existing cyber security protocols enough to be well prepared to face the rapidly growing, ever-threatening cyber warfare?

9.4 Cyber Physical System–War Gaming

"Various armed forces around the world have conducted War Games for many years to test their responses, train staff, and for purposes of operational analysis." – Dark Guest: Training Games for Cyber Warfare Volume 1 [8].

War Gaming has time and again proven to be an extremely useful instrument for the armed forces. From there, it has been taken as a supplementary tool by governments and successful businesses. The US Army Research Laboratory (ARL) Computational and Information Sciences Directorate, Network Science Division, carried out a War Game named "Terra" [9], in December 2016, to analyse their system's cyber security module. The players were assigned to attack and defend their system. ARL have utilized War Gaming strategies in their Cyber Physical System and recorded the game events and the teams' strategic plots in order to evaluate them for the improvement of their system.

"In an attempt to educate operators, owners and users of Cyber Physical Systems, the ARL recruit players for red (attack) and blue (defense) teams and conduct Cyber Physical System-War Games in an environment as close to realistic as possible." Cyber Physical System-War Games, EJM Colbert [10].

9.4.1 War Games: Advantages

The following are the advantages of adding Cyber Security System–War Gaming as a tool to the cyber security defence procedure.

1. It supplements the existing cyber security protocol system, and does not replace existing training materials.
2. It secures an immersive experience, where players can examine the options, acquire resources, and compete to achieve their objectives.
3. It ensures compatibility with the auditing requirements of the industry.
4. It engages the players and designers in a fun manner.
5. It paves the way for the company to acquire a leading edge over others in the market.
6. It ensures thorough elaborate discussions at the end of each game session, leading to a deeper understanding of the security protocols and, in turn, results in fewer mistakes.

In addition to the known Penetration tests, Vulnerability assessment, and cyber security control audits, Cyber Physical System-War Games can be used to extensively train the cyber security professionals in domain knowledge. After the risk assessment stage, it would be a comprehensive step to engage the cyber security professionals and executive authorities to undergo the War Gaming experience, in order to carry out "smart" training of the personnel to strongly establish the existing policies and procedures, and to create future plans. During the War Gaming session, security personnel can exercise their skill sets in an enjoyable and interactive fashion. Carefully collected data from the training are then used in assessing the efficiency of the existing cyber defence mechanisms. Once any operational weaknesses or security gaps are identified, this knowledge can be used to fix them, with the goal of strengthening the cyber security firewalls.

War Games, being one of the layers in the multi-layered Cyber Physical System, provide a simulated, content-rich environment to test the handling of vulnerability of a real-time system, with test attacks as close as possible to the real-time security attacks. War Gaming has the efficiency to help the corporate sector to be prepared well ahead of threats and to avert security threats in critical cyber infrastructure, while, at the same time, being a fun and innovative tool. It can be considered to be a supplement, similar to the vitamin pills supplementing our body's nutrients. War Gaming is a lesser-known utilization of gaming that has a staggering impact on the world. It is a small nook in the Cyber Physical System, but could still turn out to be an obvious but innovative way to stay ahead of the competition.

War Gaming could prove to be an essential tool that aids us in going beyond the norm in order to make sure that the in-house security procedure follows the standards, and offers an opportunity for the companies to strengthen their IT defences, in compliance with the audits. When executed periodically, War Gaming could very well strengthen the Cyber Physical System of the company, be it a large- or a small-scale operation. War Gaming can be viewed as a dramatic display of the protocols in place, that takes the documentation work of Penetration testing and gamifies it, for lack of a better term, into an important asset. It can be a comprehensive step to facilitate the cyber security professional, managers and executive authorities to undergo the War Gaming experience. The experience is a by-product of the training, in addition to establishment of the security policy pipeline, identification of operational weakness, cyber security gap discovery and the eventual cyber security design rework. A retrospective session can be held, and any required mitigation steps could be accredited systematically and added to the security protocol in place. Recognizing and acknowledging the possibility of using War Gaming to strengthen the cyber security systems will open the mind and go a long way to providing the company with a leading competitive edge in the market.

9.4.2 Architecture

Security professionals of the company, especially new employees, are normally provided with cyber security training as part of the training architecture. It would benefit the company to introduce Cyber Physical System–War Gaming, in order to complement that training package. It can be used to explore the possibilities of hacks and attacks, and very well extend it to a yearly gaming experience, where cyber security professionals, of all levels and all modules, participate with clear objectives. Periodic War Gaming procedures would help the professionals to stay in contact with rapidly growing trends in the field, as well as to go phase-by-phase through the existing cyber security protocols, and finally to devise new procedures that would add to the secure infrastructure, strengthening the existing infrastructure as an end result.

Games can be designed in such a way that they focus on the threats, ones which are critical to the day-to-day functioning of the systems. Cyber Physical System–War Gaming can be utilized to measure the capabilities of the professional during each phase of the pipeline. For example, one of them would be to measure the time taken by an analyst to figure out the location of the vulnerability. This data could be tracked, stored and analyzed by the manager, located higher up the organization hierarchy. This would be part of the multifarious data which would help to identify any lags in the process. The other data that could be tracked would be the number of clicks taken to reach the source of the vulnerability vector, which could be obtained using screen-capturing software. These lags can be mitigated by making the players aware of the protocol, to train them repeatedly in that area of requirement and to keep track of their progress. Hereby, the lag detection can be turned into a strength of the user, henceforth making the system stronger with each War Gaming iteration, and, in turn, making the whole industry stronger, iteration by iteration.

These core concepts should be kept in mind when designing Cyber Physical System–War Games:

1. Risks to pursue.
2. Third-party tools used, if any.
3. Sandboxes used during game time.
4. User interface designs that aid and abet the players.
5. Participants carefully chosen from various teams.
6. Rewards and achievements.

Several variations of War Gaming techniques are widely available, including, but not limited to, table top exercise, deductive games, two-sided games, and scenario-based games. Any of these techniques can be used to create a simulated environment for realistic evaluation and inference on unexpected uncertainties. Application domains determine the implementation details of the War Gaming system and not the other way around. For example, War Gaming for a chemical plant's Cyber Physical System could include scenarios against hacking the internal network connection or against maliciously created faults in the chemical formulas administered. On the other hand, War Gaming for an Artificial Intelligence system in the defence domain could aid in preventing attacks on the mathematical model or the decision-making cores of the AI. The players should be able to assess the critical situation, make quick decisions, and be able to dynamically act upon them in their respective application domains.

Of course, War Gaming must be practised in a non-production environment, made possible by placing debug hooks into the product. It goes without saying that the production environment product does not have these debug hooks. The Cyber Physical System–War Gaming training programme can be set up in such a way that a training mentor who knows the ins and outs of the complex system is available to guide the players through the entire gaming experience.

The war game can be designed in such a way that it focuses on one threat after the other. This is to ensure that each threat is being handled in depth by the professionals, from the time when the threat is identified to the time where it is mitigated or passed over to the respective department. The players from each department can be selectively chosen and given prior notice, so that they can be ready for the game. This is to ensure that the players are well prepared and any gaps or lags in the cyber security pipeline identified are those which go beyond the scope of preparations.

Known vulnerabilities can be made available, and the players can be subjected to the training environment where they must identify the threats, figure out the possible locations where the attack might happen, and take every possible step under their control to mitigate the threat. In case they found that the attack originated from a different module or that they found that the possible mitigation solution needs to be executed by a different team over which they do not have control, they can raise a red-flagged bug to the respective department. All the relevant details of their investigation need to be attached to the raised red-flag bug, and the level of relevant detail can be one of the metrics by which the players are assessed. Hence, it is crucial for the cyber security professionals from various modules of the product to participate in the Security–War Game. The pool of analysts from various teams participating in the War Game makes sure that the problem is handled by the respective team as soon as it is passed on to them.

These questions are best asked and documented for any tangible updates to the existing cyber security structure of the company. The take-home messages at the end of the War Gaming session should be as follows:

1. Were the responsibilities of each cyber security personnel provided to them in a well-documented manner before the commencement of the War Gaming session?
2. Did the players/participants have proper access granted to them prior to the beginning of the War Gaming session?
3. Did the trainer explain the goals and objectives of the War Game in a clear, well- documented manner?
4. Did the communication between the inter-departmental employees happen within the well-defined framework of the company's organizational structure?
5. Was the time taken by each player during each stage of the War Gaming play minimal? Was it tracked, saved and analyzed?

6. Were there any redundancies that could possibly be avoided or averted?

In the end, the whole game experience needs to be recorded, documented and discussed, for it to be fully effective.

9.5 Summary

Keeping toe-to-toe with the cyber security guidelines requires consistent re-evaluation of the threats, interpretation of the actions by the players, and vigorous tuning of the security protocol pipeline in order to cater to the guidelines.

Cyber Physical System–War Gaming is, unfortunately, a rare technique in the current cyber infrastructure, which, if introduced into the cyber security training protocol, could efficiently help the professionals and authorities to identify any gaps in the identification and deterrence process and to prevent malicious misuse. It can also help to accidentally stumble upon any unknown vulnerabilities or gotchas in the process and, in turn, to add mitigation steps for the same, in order to strengthen the system and to keep up with the rapidly growing threats. Cyber Physical System–War Gaming will prove to be a structured and fun way to manage cyber security throughout the company, while also keeping the players immensely engaged in the process. The mitigation steps would be accredited by the cyber security in-charges and added to the cyber security process.

War Gaming experiences provide a feedback loop to verify the Cyber Physical System against cyberattacks, both expected and unexpected, and, at the same time, to analyze the collected data, from which deductions could be used to enhance and adapt the system into a better, more resilient one.

References

1. Kriegsspiel, Wikipedia, https://en.wikipedia.org/wiki/Kriegsspiel.
2. *Simulating War, Studying Conflict Through Simulation Games, Philip Sabin,* Bloomsbury Publishing India, 2012.
3. Reality Is Broken: Why Games Make Us Better and How They Can Change the World, Jane McGonigal, Interactive 2011, South By SouthWest.
4. India Among Top 5 Markets for Mobile Gaming, Gaurav Laghate, *Economic Times,* https://economictimes.indiatimes.com/tech/software/india-among-top-5-markets-for-mobile-gaming/articleshow/65396386.cms?from=mdr.

5. Gamification in Business and Education - Project of Gamified Course for University Students, Michal Jakubowski Kozminski University.
6. Gartner Redefines Gamification, Brian Burke 2014.
7. Simulating a Super Brain: Artificial Intelligence in Wargames. Sydney J. Freedberg. Jr. https://breakingdefense.com/2019/04/simulating-a-super-brain-artificial-intelligence-in-wargames/.
8. Dark Guest, Training Games for Cyber Warfare, Volume 1, Wargaming Internet Based Attacks, John Curry and Tim Price MBE, 1.
9. Edward J Colbert et al. 2017. *Terra Defender Cyber-Physical Wargame*, https://pdfs.semanticscholar.org/b5f7/86c6c7f698cf472cfd5baaf00275df94b465.pdf?_ga=2.225512344.300702528.1598001390-1395500756.1598001390
10. Cyber-Physical War Gaming, EJM Colbert, DT Sullivan, A Kott, 2017, https://arxiv.org/ftp/arxiv/papers/1708/1708.07424.pdf

10

Blockchain-Enabled Cyber Physical System – A Case Study: SmartGym

Keerthivasan Manavalan, Rithish kesav Saravanan, Venkatachalam Thiruppathi, Roshan Rajkumar and Kishoram Balaji

CONTENTS

Organization

Section 1 deals with terms and terminologies for the readers to understand the chapter. **Section 2** describes the SmartGym as a Cyber Physical System. **Section 3** elucidates the need for Blockchain in SmartGym Cyber Physical Systems and provides the technology overview. **Section 4** describes platforms for Blockchain. **Section 5** deals with protecting the Cyber Physical System with a decentralized Blockchain. **Section 6** gives an overview of the SmartGym ecosystem and portrays the futuristic conceptual layered architecture for Blockchain implementation. **Section 7** concludes the chapter, with comments on how to make SmartGyms smarter.

10.1 Terms and Terminologies

This section describes certain terms and terminologies for the convenience of the reader to understand the rest of the chapter.

Internet of Things (IoT): A system of uniquely identifiable, interconnected, computational, mechanical and digital machines with an ability to transmit data over a network.

Blockchain: Time-stamped sequence of immutable records managed by a group of computers, not possessed by any single entity. Each of these blocks is secured, bound to each other by cryptographic principles, to form a chain.

Distributed Ledger: It is a consensually shared and synchronized database, existing across multiple sites, that allow transactions to have public witnesses.

Mining: In the context of Blockchain technology, mining refers to the act of adding transactions to a distributed ledger of existing transactions, known as the Blockchain.

Consensus Protocol: A set of rules that defines how the nodes in a network agree on the state of Blockchain, thus making it a self-auditing ecosystem.

Smart Contract: A protocol intended to enforce, verify and facilitate the proceedings of an agreement which recognizes and governs the participants of the Blockchain network.

Ethereum: A Blockchain-based platform, featuring smart contract functionality.

Ether: Computational services and transaction fees of the Ethereum network are paid using the fundamental token called the ether.

Hash: A function that creates an encrypted output of fixed length from an input of letters and numbers (Blockchain transactions), primarily used for ensuring integrity of transmitted data.

10.2 SmartGym

The Internet of Things (IoT) refers to the existence of a connection between Cyber Physical Things for data transfer. Since the simple hardware does not have the ability to connect, efforts are taken to augment the hardware with software; together, it is simple to network the hardware and it is called IoT. Cyber Physical System (CPS) forms the first level and IoT forms the second level of digital amalgamation. IoT platforms realize centralized architectures for device discovery, processing, integration, management, notification and real-time analytics. In addition to integrating services, there are many circumstances that need autonomous communication between smart devices without the need for a central server. Decentralized peer-to-peer (P2P) models can help to empower the capabilities, such as messaging, datasharing, maintaining a trusted relationship among peers and the like. Realizing such decentralized capabilities requires an infrastructure to perform autonomously without a centralized authority. Blockchain offers a mechanism to enable such a distributed model. If the idea is harnessed to industrial automation, it is referred to as Industrial Internet of Things (IIoT), Cyber Physical Production Systems (CPPS) or Smart Factory, etc. If the idea is extended to equipment in gyms, such as light, fans, steamers, treadmills, cycles, cross-trainers, bands and the like, the gym becomes a 'SmartGym', as shown in Figure 10.1.

For the realization of an automated gym (a level ahead of currently existing gyms, in terms of customization), an integration of IoT and Blockchain technology is required, which results in the creation of a Cyber Physical System – a SmartGym. Consider a SmartGym with intelligent exercise machines, that

FIGURE 10.1
SmartGym backend processing and services for Internet of Things (intelligent machines)

perform the task of recording the workouts made, producing data, undergoing preventive maintenance, calculating statistics, counselling clients on diet, recommending a balanced intake, etc. A digital twin of this gym equipment is software that gives the output as the 'states' of this machine, in a similar pattern to that followed by the actual machine. From the SmartGym scenario, it is clear that the Cyber Physical System is a self-contained entity system, which consists of a physical thing and its digital twin, connected together. The digital twin, SmartGym equipment, virtually repeats the behaviour of the physical machine, that gives a response with improved readability when stimulated for actions and is realized using sensors and actuators.

10.3 Blockchain Technology

A SmartGym inherently has an intensive interconnectivity and a computational platform and is vital for real-world implementation of CPPSs. Blockchain empowers the Cyber Physical Systems to enhance security and bring transparency to the ecosystems. Blockchain offers a scalable and decentralized environment to gym equipment, platforms, and applications. In an interconnected world, cryptocurrencies are turning into an undeniably appealing option for developing markets, particularly those that might not have a conventional financial framework. In order to have a secure transaction, Bitcoins and the other cryptographic forms of money depend on a type of database that can keep a reasonable track of enormous volumes of exchanges safely. The innovation of the Blockchain technology gives one of the most proficient arrangements, used by a large portion of the world's largest digital currencies. This technology has turned into worldwide news, with the development of cryptographic forms of money, like Bitcoin. Currently, this innovation is impacting all business sectors and changing the manner of everyday business. Indeed, the Blockchain innovation is changing our reality. The Blockchain is a computerized record of exchanges that can be customized to record practically everything of importance. Each rundown of record in a Blockchain is called a block. Therefore, a Blockchain is an endlessly developing rundown of records, or blocks, which are safely connected.

The regular Blockchain framework consists of two kinds of records, namely transactions and blocks. Transactions are the activities completed in a specific period and are put together in a block. The Blockchain is exceptional as each block contains a cryptographic hash that connects them to the previous transactions, making a minimal series of transactions. The participants can see the transactions and can check or reject transactions, using consensus algorithms. The affirmed information is recorded into the record as a major aspect of a transaction and verified through cryptography. These chains are difficult to predict, which makes it simple to detect any alterations.

10.3.1 Hashing Algorithm

Hashing alludes to the idea of taking a discretionary measure of input data, applying an algorithm to it and creating fixed-size output data called the hashed output (digested message), which functions as the core of the Blockchain technology. The frequently used hash algorithms are of two types, namely SHA (Special Hashing Algorithm) and MD5 (Message Digest). Hashing means taking into account an input string of any length and giving out an output of a fixed length. With regard to cryptographic forms of money, like Bitcoin, the transactions are considered to be an input and passes through a hashing algorithm (Bitcoin utilizes SHA-256), which gives an output of a fixed length.

10.3.2 Blockchain Protocols

The Blockchain does not require any outsider to do transactions on his/her behalf. The network has to have the consensus mechanism. This mechanism is implemented by the Blockchain, whereas the strength of the network is also determined by the Blockchain. To maintain coherence of the data among the participating nodes of the network is the objective of the consensus mechanism. It aims to eliminate two known problems with digital currency and also to eliminate the Byzantine Generals problem, from computer science, where the parties must agree on a single strategy to avoid complete failure, but where at least one of the parties is considered to be unreliable. Some of the key algorithms are further developed to suit various applications as a part of the Blockchain protocol.

10.3.2.1 Proof of Work

As required by the Blockchain protocol, brute force is used to solve cryptographic puzzles for all the nodes on the network. Based on its output, the transactions are tentatively committed. At specific synchronization intervals, the winning node, which is created by a selected block, is broadcast to all the nodes. Using peer-to-peer (P2P) communication, the block is transmitted to all the other nodes and is included in the Blockchain. If any tentative transactions appear, they are rolled back. The agreement is accomplished as 51% power as opposed to 51% of individuals check. All other nodes, except the winning nodes that use the computing power, are wasted.

10.3.2.2 Proof of Stake

Excessive computations, implemented for Ethereum and certain Bitcoins, are not relied on by the stake protocol of block verifications. A mining power exists in the splitting of blocks across the relative hash. The stake blocks are split proportionally to the current wealth of the miners as a proof of stake protocols. The thinking behind the proof of stake is that it might be progressively harder for miners to procure large enough amounts of digital currency cash than to acquire suitably incredible computing equipment. It is a vitality sparing option.

10.3.3 Advantages of Blockchain

The following are some of the advantages of the Blockchain technology, a technology that promises to revolutionize finance and business:

- **Minimizes cost:** As Blockchain sets up a P2P network within one system, it minimizes the time and expense of intermediaries, such as the middlemen.
- **Fast and convenient:** It is complex to use different ledgers and processes throughout the lifecycle of a transaction, whereas a stock purchase transacted in a Blockchain settles in a minute. Another entity is not required to process the transaction *via* Blockchain.
- **Secure:** Every single transaction is stored in a block that connects to the ones before and after it, which amplifies the security. Although nothing is hackproof, the Blockchain is much more secure than all other systems today.
- **Transparent and incorruptible:** The transactions are truly transparent and easily verifiable. The Blockchain data is not stored centrally, so, as it is decentralized, it makes it difficult for any hacker to corrupt.

10.4 Platforms for Blockchain

This section explores various Blockchain platforms and their purpose. This will help the readers to choose their platform, based on their requirements.

10.4.1 Comparisons of Blockchain Platforms

Ethereum is an open software platform that enables designers to design, develop and deploy decentralized applications. Hyperledger Fabric is a framework with plug-and-play components for developing Blockchain-based products, solutions and applications, for use within private enterprises. Hyperledger Sawtooth is an enterprise Blockchain platform for building distributed ledger applications and networks. Hedera Hashgraph accelerates distributed ledger technologies (DLT) to transform existing markets in various industries. Ripple uses a unique distributed consensus mechanism through a network of servers to validate transactions. By conducting a poll, the servers or nodes on the network decide, by consensus, about the validity and authenticity of the transaction. Quorum is an Ethereum-based distributed ledger that combines the innovation of the public Ethereum community with enhancements to support enterprise needs. Hyperledger Iroha is a specially designed distributed ledger technology for infrastructural/IoT projects. In a permission-less Blockchain platform, all data are shared with all parties and are largely unsuited for businesses. Corda is flexible and scalable, ensuring the highest level of privacy and security by sharing data with only relevant parties. EOS Blockchain architecture is designed to facilitate both vertical and horizontal scaling of decentralized applications, supporting accounts, authentication, databases, asynchronous communication and the scheduling of applications across multiple Central Processing Unit (CPU) cores and/or clusters. OpenChain distributed ledger technology is suited for organizations needing to issue and manage digital assets in a robust, secure and scalable way. Stellar is an open source, decentralized protocol for digital currency, used to approve cross-border money transactions between any pair of currencies (Figure 10.2).

10.5 Protecting CPS with Decentralized Blockchain

Current-day Cyber Physical Systems (CPS) have faith in centralized, middleware-based communication, ranging from thin-client fat-server to fat-client thin-server models. Physical devices in CPS are connected through cloud data centres with support for big data processing and storage. Internet-based connectivity between devices is seen as a model for IoT systems with centralized architecture. A decentralized technique to deal with CPS requires a standardized peer-to-peer communication model, wherein everyone shares equal stakes and benefits. Also, computation and storage needs are distributed across the devices that form networks. This will prevent single-point failure in a network, avoiding total failure. Blockchain provides a brilliant solution, which permits the creation of a distributed digital ledger of transactions. These are shared among the nodes of a network, as an alternative to being stored on a central server. Furthermore, it addresses the issues of conflict of authority among the stakeholders, by providing a secured platform.

10.6 Gym Ecosystem

Gym equipment consists of sensors and actuators to convert and transmit the data to its intended computing storage devices. A sensor in the gym

If requirement is

(

Public Blockchain network, Open-source platform for decentralized applicationsand accessible anywhere in the world.

)

then using [8], recommendation is 'Ethereum'

Else if requirement is

(

Permissioned distributed ledger framework for broad range of industry use cases

)

then using [9], recommendation is ' Hyperledger Fabric'

Else if requirement is

(

Enterprise solution for developing transaction-based updates synchronized by consensus algorithms to share between untrusted parties.

)

then using [11], recommendation is 'Hyperledger Sawtooth'

Else if requirement is

(

decentralized public network ; create fast, fair, and secure applications

)

then using [12], recommendation is 'Hedera Hashgraph '

Else if requirement is

FIGURE 10.2
Logical Platform, with A. positive, B. negative and C. neutral responses

equipment changes a physical parameter to an electrical output, whereas an actuator converts an electrical signal to a physical output. The sensor is situated at the input port to take the input, whereas an actuator is placed at the output port. For example, proximity sensors are used in treadmills for speed and gradient sensing. The speed of the belt is determined by positioning a

```
( develop real-time settlement system, currency exchange, and schematics.
for free global financial transactions )
            then using [15], recommendation is ' Ripple'

Else if requirement is
(
combine the innovation of the public Ethereum community for enterprise needs
)
            then using [13], recommendation is 'Quorum '
Else if requirement is
(
Unique chain-based Byzantine Fault Tolerant consensus algorithm, called Yet Another
Consensus and the BFT ordering service.
)
            then using [10], recommendation is 'Hyperledger Iroha '

Else if requirement is
(
transact directly and ensure strict privacy using smart contracts, reduce transaction and
record-keeping costs.
)
            then using [6], recommendation is 'Corda '

Else if requirement is
```

FIGURE 10.2
(Continued)

reed sensor near the drive belt. Belt tensions are monitored and kept adjusted. Apart from speed sensing, the gradient of the treadmill is adjusted by raising the front wheels. Locating reed sensors in a very small number of strategic positions on the arms of the front wheels can provide information to the controller about the gradient. The uninterrupted information, collected

```
(
Facilitate vertical and horizontal scaling of decentralized applications , schedule applications
across multiple CPU cores and/or clusters
)
            then using [7], recommendation is 'EOS '
Else if requirement is
(
distributed ledger technology to issue ,manage digital assets in a secure and scalable way.
)
            then using [14], recommendation is 'OpenChain'
Else if requirement is
(
Decentralized protocol for digital currency to approve money transfers across cross-border
transactions between any pair of currencies.
)
            then using [16], recommendation is 'Stellar '
```

FIGURE 10.2
(Continued)

by the sensors, is then used to save health information to a mobile device or a website. Other examples of sensors include a heartrate sensor, velocity sensor, etc. A SmartGym ecosystem, using IoT application, is one wherein gym equipment is connected to various tracking devices which measure and record real-time data to optimize fitness performance through diet and exercise feedback. The exercise feedback is given to the user *via* fitness bands or mobile applications.

Mapping of a gym user's environment is the key central to implementing Blockchain technology in a SmartGym ecosystem with IoT (Figure 10.3). Figure 10.4 shows a typical gym environment from a gym user's perspective. This helps us to develop a user-friendly Blockchain-implemented SmartGym environment, using IoT applications. To enable message exchanges, IoT devices will leverage smart contracts which then model the agreement between the two parties. It answers the challenges of scalability, single-point

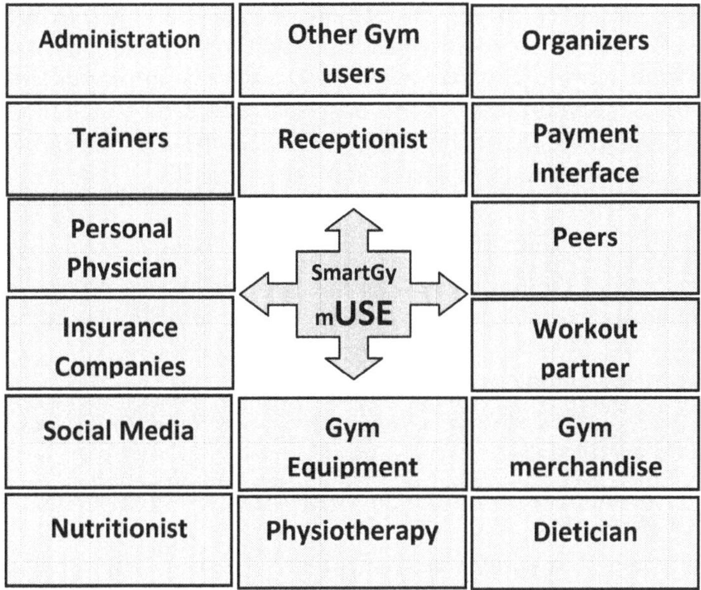

FIGURE 10.3
SmartGym user environment

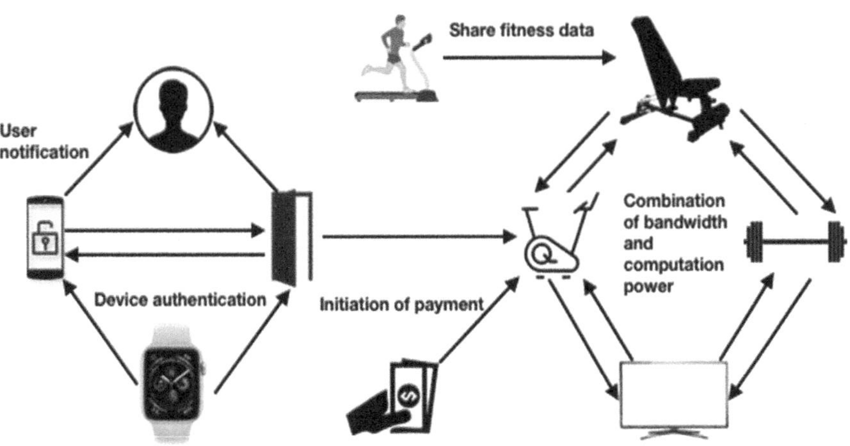

FIGURE 10.4
Typical gym environment, from a SmartGym user's perspective

failure, time stamping, record, privacy, trust and reliability in a very consistent way. In this case, IoT data and Blockchain can lead to intelligent automated gym data records and policy applications. Although this might seem like a very futuristic application of Blockchain to the gym industry, it is worth noting that, from a gym user's perspective, a more decentralized flow and storage of data and records is to be preferred for both sharing and developing. However, these are not the only applications that fit into the scope of IoT. But having a wider picture of Blockchain technology can make a SmartGym smarter. The even bigger picture is the implementation of AI along with Blockchain in IoT, with deep learning algorithms, next-generation security of data, edge computing and fog computing, high-performance embedded chips, computer vision and machine vision.

10.6.1 Blockchain Implementation

Ethereum, one of the popular Blockchain platforms, has been adapted for prototyping the SmartGym. Ethereum is popular for its following features:

- It is Turing-complete.
- It is a replicated state machine and every node knows everything, called 'world state'.
- Smart contract is a code written in an Ethereum-specified language, such as Solidity.
- It is immutable, which means that there is resistance to change, once deployed.
- It is deterministic since, for a given input and present state, the output of the program is the same for everyone.
- It is virtualized, as Ethereum runs on virtual machines.
- It is decentralized, because a copy of it runs on every Ethereum machine.

The evolution started in 2011, when Vitalik Buterin co-founded Bitcoin magazine with Mihai Alisie, and in 2012 Buterin published the Ethereum White Paper. In 2014, official Ethereum announcements were made, and Gavin Wood published the Ethereum Yellow Paper, and a foundation was setup in Switzerland. A white paper is a marketing document, used to persuade potential customers to use a particular service or technology, like a proposal. A yellow paper, on the other hand, is a more technical version of the white paper. It presents the scientific details of the technology in a very concise way. From 2015 to 2018, Ethereum versions 1.0 to 3.0 were released.

SmartGym tracks the equipment, collects users' health data and stores them in Blockchain. Blockchain is preferred because it can be used to prevent forging or modifying of an individual's workout data. The gym equipment devices exchange data through a Blockchain to establish trust among themselves, instead of going through a third party, which obviously reduces the operation cost. The distributed ledger technology allows for a list of interactions between the gym equipment devices. It is possible to keep track of devices that interact, and also of changes made during the interaction. For example, changes could occur with respect to the heartbeat rate of the user or the calories burned or to monitor whether the specific daily workout target was reached. All of these can be stored in the cloud too and accessed, whenever needed. These data are stored in an encrypted form. Many companies are involved in the implementation of SmartGym projects, using Blockchain technology. Furthermore, Blockchain has a high reliability as it is a decentralized ecosystem, protecting the data from gym devices from unauthorized alterations. Blockchain enables device autonomy *via* smart contract, individual identity and integrity of data, and supports P2P communication. Figure 10.5 shows the steps involved in setting up of the Blockchain environment.

Ganache is a personal Blockchain for Ethereum development, to deploy contracts, develop applications, run tests and inspect state, while controlling how the chain operates. In the main Ganache window, there exist addresses with a balance of 100 ethers (ETH) each. Ganache provides a personal Blockchain to start immediately. Node.js is a platform building fast and scalable network applications. Node.js uses an event-driven, non-blocking input/output (I/O) model that makes it lightweight and efficient, perfect for data-intensive real-time applications that run across distributed devices. Truffle is a developer environment, testing framework and asset pipeline for Blockchain. It allows developers to spin up a smart contract project and provides project structure, files and directories that make deployment and testing much easier.

10.6.2 Layered Architecture for Blockchain-Enabled SmartGym

The conceptual architecture for SmartGym can be viewed as a layered architecture, as shown in Figure 10.6, where each layer is separated from other layers so that contributors can augment patches to any new layer with both physical and logical data independence, without affecting the SmartGym ecosystem.

- Layer 4: **Physical layer**, designed to focus on internal capabilities in the context of storage, such as using the cloud.
- Layer 3: Provides services for **network and communication**. The services include data and equipment security services, equipment

Step1:Install Ganache

Step 2:Install Nodes and NPM

Step 3: Install Truffle and create a new truffle project

Step 4: Using Truffle console, configure Truffle to connect to Ganache.

FIGURE 10.5
Setting up of a Blockchain environment

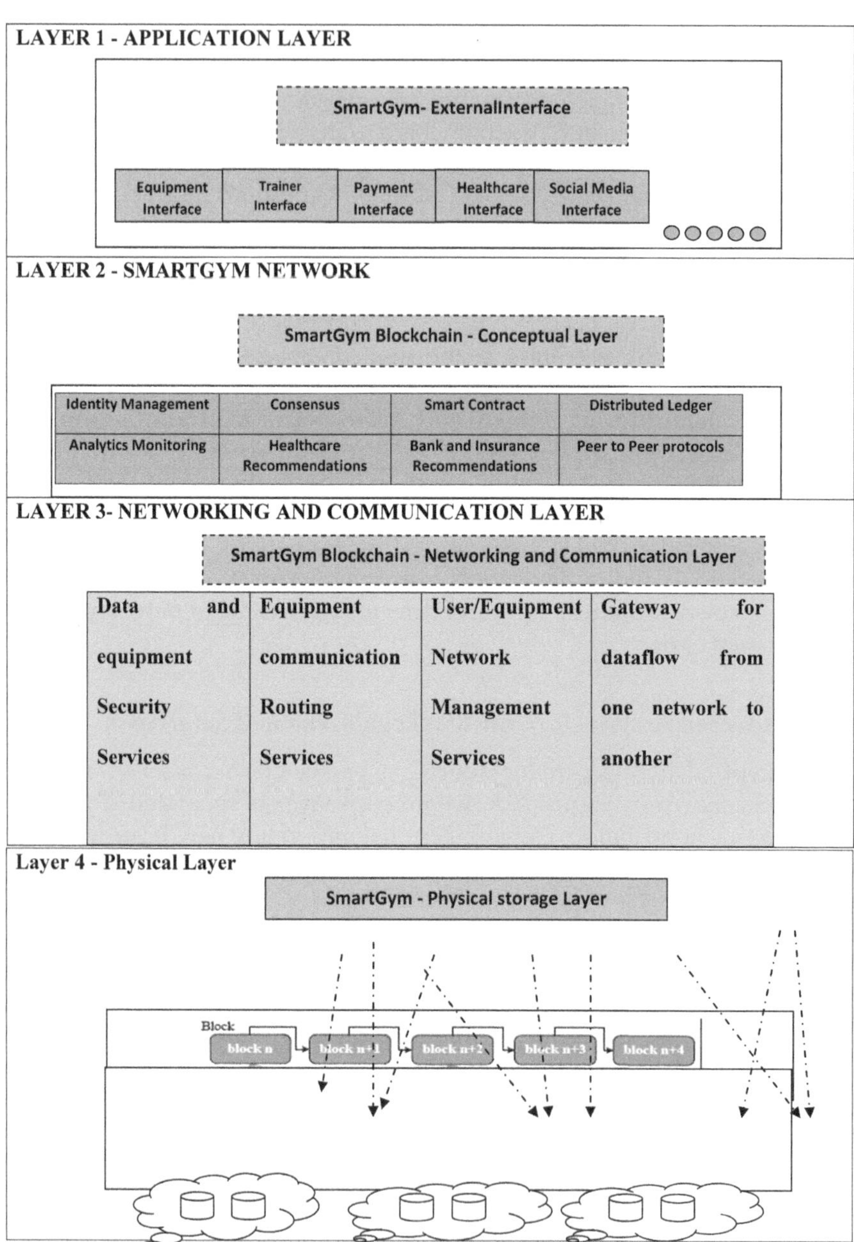

FIGURE 10.6
SmartGym layered architecture

communication routing services, user/equipment network management services and gateway for dataflow from one network to another.

- Layer 2: A **conceptual layer** that aids in features such as identity management, smart contract, analytics monitoring, consensus and the like.
- Layer 1: Designed as an **application layer,** that deals with external interfaces to SmartGym, such as interfaces between equipment, trainers, payment and the like.

10.7 Conclusion – Making Smart Gyms Smarter

With the proliferation of connected equipment in modern-day gyms, there is a need to address the challenges to scalability and security. Connected items of equipment are heterogeneous, with different manufacturers, so that distinguishability and interoperability need to be guaranteed in a secure way. The current gyms are increasingly using wireless sensors, that help members monitor their work rate. The instructors are familiar with the data, which, in turn, help them to understand the gym user and serve the needs of various group members. Shared tracking has become crucial for people as they get to share their experience with family and friends, as such activities motivate both the user and their contacts. Statistics report that over 30% of the users abandon the use of wearable trackers as they get bored eventually, if their data cannot be shared or if they do not receive feedback on how they work out. Blockchain developers have to take the psychological characteristics into account in the concept of app development. Furthermore, augmenting machine-to-machine (M2M) protocols can improve the performance, stability, flexibility, end-to-end security of real-time data and cross-platform interoperability.

Bibliography

1. Atlam, H.F.; Alenezi, A.; Alassafi, M.O.; Wills, G.B. Blockchain with Internet of things: Benefits, challenges, and future directions. *Int. J. Intell. Syst. Appl.* 2018, *10*(6), 40–48.
2. Conoscenti, M.; Vetro, A.; Martin, J.C.D. Blockchain for the Internet of Things: A systematic literature Review. In: *Proceedings of the 3rd International Symposium on Internet of Things: Systems, Management, and Security, IOTSMS-2016,* Agadir, Morocco, 29 November–2 December 2016.

3. Özyılma, K.R.; Yurdakul, A. Integrating low-power IoT devices to a blockchain-based infrastructure. In: *Proceeding of the Thirteenth ACM International Conference on Embedded Software 2017 (EMSOFT '17)*, Seoul, Korea, 15–20 October 2017. doi:10.1145/3125503.3125628.
4. Samaniego, Mayra; Deters, Ralph Blockchain as a service for IoT. In: *IEEE International Conference on Internet of Things (iThings) and IEEE Green Computing and Communications (GreenCom) and IEEE Cyber, Physical and Social Computing (CPSCom) and IEEE Smart Data (SmartData)*, IEEE, Chengdu, China, 2016.
5. Zyskind, G.; Nathan, O.; Pentland, A. Enigma: Decentralized Computation Platform with Guaranteed Privacy. *arXiv Preprint ArXiv:1506.03471*.
6. https://www.corda.net/.
7. https://eos.io/.
8. https://www.ethereum.org.
9. https://www.hyperledger.org/projects/fabric.
10. https://www.hyperledger.org/projects/iroha.
11. https://www.hyperledger.org/projects/sawtooth.
12. https://www.hedera.com/.
13. https://medium.com/blockchain-at-berkeley/introduction-to-quorum-blockchain-for-the-financial-sector-58813f84e88.
14. https://www.openchain.org/.
15. https://www.ripple.com.
16. https://www.stellar.org/.
17. https://github.com/trufflesuite/ganache/releases.

Section 3

Advances, Challenges and Opportunities in Cyber Physical Intelligence

11

Context-Aware Computing for CPS

Bhuvaneswari Arunagiri and Maheswari Subburaj

CONTENTS

Organization of the Chapter

Section 1 presents the terms and terminologies for the user to understand the chapter. **Section 2** discusses the layered architecture and the attributes of the Cyber Physical System. **Section 3** explains the context-awareness in the Cyber Physical System (CPS) and provides an example. **Section 4** deals with the sensors and their implementation, using the Generic Sensor APIs. **Section 5** provides an insight into the Semantic Sensor Network Ontology, which is used to relate the shared concepts in CPS. **Section 6** explains about CPS in the healthcare domain. **Section 7** proposes the system architecture for Context-Aware Healthcare Cyber Physical System (CA-HCPS). Finally, **Section 8** discusses the open challenges and research opportunities in CPS for the healthcare domain.

11.1 Terms and Terminologies

This section defines the most important terminologies used in this chapter.

Cyber Physical Systems (CPS): CPS combine sensing, computation, control and networking into physical objects and infrastructure, connecting them to the Internet and to each other.

Sensors: A Sensor is a device which detects or measures a physical property and records, indicates or otherwise responds to it.

Context-Aware Computing: Context-awareness is the ability of a system or system component to gather information about its environment at any given time and adapt behaviours accordingly. Contextual or context-aware computing uses software and hardware to automatically collect and analyze data to guide responses.

Sensor APIs: The Sensor APIs are a set of interfaces built to a common design, that expose device sensors in a way consistent to the Web platform. The Generic Sensor API consists of the base sensor interface, with a set of concrete sensor classes built on top, such as accelerometer, linear acceleration and sensor, or gyroscope, absolute orientation sensor and relative orientation sensor.

Ontology: Ontology is a "formal, explicit specification of a shared conceptualization". Ontology provides a shared vocabulary, which can be used to model a domain that is the type of objects and/or concepts that exist, and their properties and relationships.

Semantic Sensor Ontology: The Semantic Sensor Network (SSN) ontology is an ontology for describing sensors and their observations, the involved procedures, the studied features of interest, the samples used to do so, and the observed properties, as well as actuators.

Web Service Repository: The Web Services Repository has details about Web Services and their definitions are imported, using Service Manager. The WSDL files that have these definitions of Web Services are stored, together with their related XML Schemas. Clients of the Web Service can then search the repository for the WSDL file, which they can use to build and send messages to the Web Service, using the Enterprise Gateway.

Atomic Services: An Atomic service is a single Web service that has WSDL as the interface to connect with other services.

Composite Services: A composite service is a Web service that has either atomic services or composite services as sub-services and the workflow between sub-services.

11.2 Introduction

The first industrial revolution happened with the mechanization of the world, which moved production from home to industry. The second industrial revolution applied science to communication, mass production and manufacturing, which improved productivity. The third industrial revolution was digitization, which started with the invention of the calculator, which was further groomed by electronics and IT systems. Currently, the world is undergoing the fourth industrial revolution, where machines are enabled, with sensors and wireless connections, to make decisions on their own. As with the other industrial revolutions, this fourth revolution also brings many changes in industry and in individuals.

The new trend focuses on automation and data analytics of physical processes, that are enriched by the advent of technologies, which include Cloud Computing, Artificial Intelligence, Cognitive Computing, Internet of Things, and Cyber Physical Systems (CPS). Industry 4.0 promotes the "Smart Industry", where the Cyber Physical Systems establish communication on top of the Internet of Things, simulate the real world, manipulate physical processes and make decisions to proceed further.

The "Smart" world depends on multiple computer science streams such as sensors, communication, networking and software engineering. The research on CPS is still in its infancy, with various challenges that demand the attention of researchers [3, 5, 6]. Such challenges include the need for standardized abstraction and architecture, seamless integration of software and hardware components, standard protocols for distributed computation and standard tools and techniques for verification and validation of the novel system. The research focus encourages multi-disciplinary collaboration between industry and academia. The research on CPS has expanded as it enhances the performance of the overall system, compared with the assimilation of software systems and sensor networks

The CPS includes sensors and actuators, that control and monitor the physical processes used for computation. Cyber Physical Systems are created by integrating computational components, networking and physical methods for a particular purpose. CPS is an extension of the existing embedded systems, such as remote cars, battery-operated toys and digital thermometers, which do not involve computation. There are many application areas for the CPS, including a Smart Grid System, Smart Healthcare, Smart Transportation, Automatic Pilot Avionics and Smart Humanoids, to mention but a few.

CPS has numerous applications in healthcare [4] as the latter depends on interconnected medical devices and the necessity for precise decisions, based on the data obtained from those devices. The intelligent hospitals, surgery by robots, decisions on drugs and therapy, remote monitoring of patients and assistance to older people are various research areas in the domains of medicine and well-being. While monitoring the patient's health, new technology is required to identify biological conditions, communicate them through devices, analyze them and suggest changes, if needed, for patient treatment. Such a machine-dependent environment for health monitoring improves with context-awareness for the benefit of the individual.

Context-aware computing is defined as "an application's ability to adapt to changing circumstances and respond according to the context of use" [8]. The term "context" may be considered as any information that describes the situation of the entity. Continuity in health monitoring is essential, which will be complete when context-awareness is included in the system. The various aspects of context are (i) user profile, (ii) environment profile, (iii) resource profile and (iv) time. The user profile will consist of the location and availability of nearby assistance; the environment profile consists of temperature, noise and traffic conditions; and the resource profile consists of nearby devices, network connectivity, communication cost and bandwidth.

It is clear that context symbolizes the meta-data available to represent the entity, which may be a person, place or thing. The recent illustration of metadata is greatly influenced by ontological representation. The ontology is a formal, explicit specification of a shared conceptualization. The conceptualization is an abstract, simplified view of the domain of interest. For example, the healthcare domain represents patient, doctor, hospital, drug and price as concepts. These concepts may be shared between two different domains such as healthcare and e-commerce, as shown in Figure 11.1. The most popular language that helps to create and manipulate ontology is the Web Ontology Language (OWL) [9].

In terms of continuous monitoring, the communication among the entities, mainly patient and doctor, is a major requirement, that is satisfied by sensors.

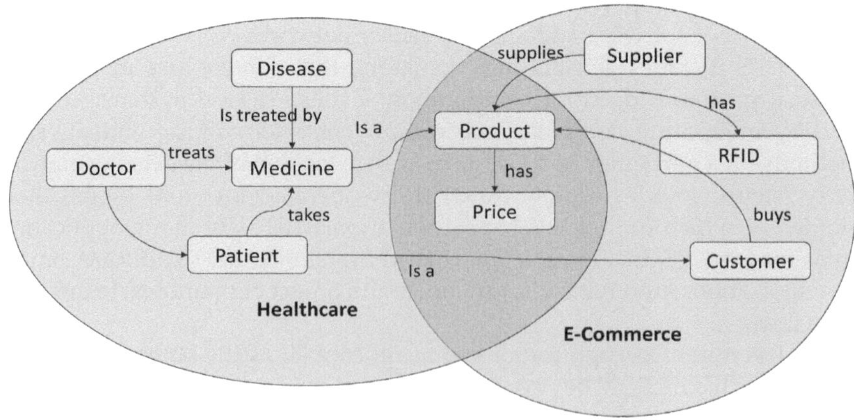

FIGURE 11.1
Shared conceptualization in ontology

For example, sensors measure the clinical condition of a patient, by sensing the temperature and other vital signs of the patient at regular intervals. The data sensed from the patients must be conveyed to the respective medical personnel, for example, the doctor, irrespective of their location. Hence, the sensed data must be exposed to the Web, which can be communicated over smart phones to reach the doctor. Ongoing research by W3C has introduced Generic Sensor API [10], intended to promote consistency across sensor APIs, and make it easier and faster to expose new sensors over the Web. It defines a blueprint for writing specifications of concrete sensors and also defines an abstract sensor interface that users can extend to accommodate different types of sensors.

Searching, reusing, integrating and interpreting the sensor data is as important as publishing over the Web. Enabling all these operations on sensor data requires greater efforts [11]. Hence, OGC's Sensor Web Enablement annotates sensors and their observations by using the standards (i) Observation and Measurement (O and M) for observation-centric approaches and (ii) Sensor ML for sensor-centric approaches. The growth of Web of Things, smart cities, smart homes, etc. showcases the importance of describing the sensors, observations, procedures, features of interest, actuators and actuations. The sensors should be extended for social sensing. Hence, Semantic Sensor Network Ontology (SSN) and Sensor, Observation, Sample and Actuator Ontology (SOSA) have been introduced by W3C [10]. The Semantic Sensor Network (SSN) ontology [10] is a shared conceptualization to express the details about the sensors and the sensed data, the related functions, characteristics and properties of sensors and related samples. Similar to sensors, the details of actuators are also included in the ontology. SOSA provides a lightweight core for SSN to broaden the users of semantic Web ontologies.

The main component of Industry 4.0 is Cyber Physical Systems (CPS), which is expected to deliver critical services by sensing the physical environment and adapting to the environment. CPS combine the cyber capabilities with physical capabilities to solve problems, which neither part can solve alone. The cyber capabilities include computation and communication and the physical capabilities include the sensors and actuators. The networked embedded devices are controlling and monitoring the physical processes, which influence the computation. The CPS applications sense the physical environment and provide the right service at the right time in any critical situation.

Figure 11.2 portrays the five layers of the Cyber Physical Systems, namely Sensor, Data Processing, Virtual, Reasoning and Configuration, from bottom to top, respectively. The Sensor layer speaks about the smart connection among the different types of sensor that are contained in the self-connecting and self-sensing devices. The Data Processing layer concentrates on transforming data into information suitable for various analytics. It obtains the data from the self-connected sensors, converts them into the required format and uses them for prediction. The Virtual layer handles the data by

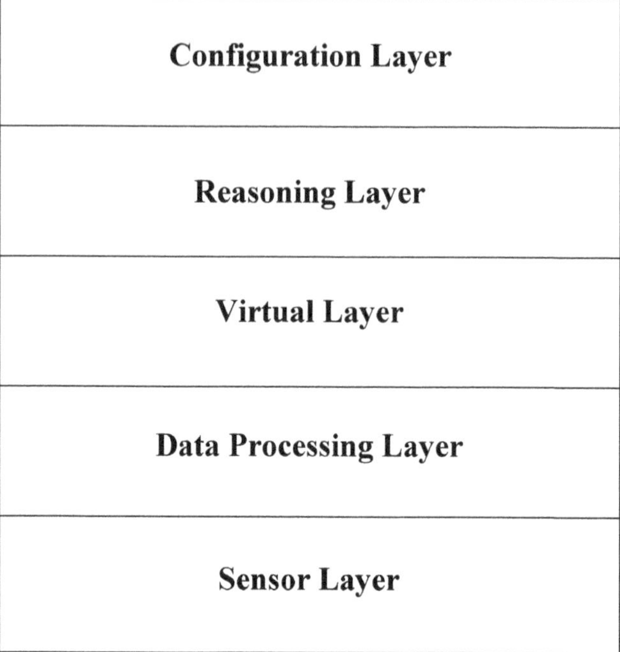

FIGURE 11.2
Layers of CPS

clustering, and provides communication between components and machine. The Reasoning layer takes care of the visualization of data obtained from self-assessment and self-evaluation of the sensed data, which helps indecision making at the critical time. The Configuration layer provides the self-adaptive CPS, which becomes reconfigured, based on the precedence and threat criteria to attain robust performance.

The CPS is characterized by the following features given below:

- **Integration**: The integration of the Sensor network and the Web is an essential characteristic in Cyber Physical System. CPS provides access control, network coordination and transactions over the network and fault tolerance, which are the primary concerns of any networked system.
- **Human-System Interaction**: It is critical to make decisions by prototyping and determining the human insight about the system because the system may be affected by the conditions that prevail in state and system-dependent variables.
- **Uncertainty**: To avoid uncertainty, there should be a process, to validate the design and measure the trust on design, with a strong proof.

- **Performance of CPS**: The CPS is expected to provide a better performance by allowing automatic re-design, based on feedback obtained from the sensors and cyber infrastructure. The other measures that characterize the CPS are scalability, flexibility and quick response time, bandwidth utilization and fault detection

Cyber Physical Systems can be approached by the SOA approach as described [2]. In the lightweight Service Oriented Architecture (SOA) approach, the devices are expected to publish their abilities as services onto a service registry. Such an approach provides service discovery, with the help of an ontology that describes the details of the sensors, such as model number, message type and scope of the sensor. The sensors are considered to be entities that can be instantiated and interconnected to gather data.

The Cyber Physical Systems are expected to interpret the sensed data with respect to the domain-specific application, maintain the relationship among the sensors and determine how these sensors can work together to improve the quality of the services related to the domain. This chapter visualizes the CPS as an abstraction of software components, that are managed by RESTful Web services. Section 11.3 deals with the context-awareness, that needs to be addressed in the Cyber Physical Systems.

11.3 Context-Aware Computing

Human-Computer Interaction understands the user and the context of the user and creates a design that aids the foreseen situations of use, i.e., a decision is taken at design time. But context-aware computing decides when the user interacts with the application, i.e., at run time. [1] refers "Context" as any information that can be used to characterize the situation of an entity. An entity is a person, place or object that is considered relevant to the interaction between a user and an application, including the user and its application.

The context-aware computing concept was described [7, 8] as the one that adjusts according to the position of use, a group of neighbouring people, devices that are reachable and also the changes to such things. The system inspects the situation and reacts accordingly. The fundamental thought is that portable devices can provide various services in diverse contexts, where the context is strongly associated with the position of a device. However, context is linked not only with location; it also includes other entities such as time, temperature and user identity. These facts can be used for both tagging information and for activating alarms.

[7] classified the context into two types, namely personal and environmental contexts. The personal context includes the profile and preferences of the

user, whereas the environmental context includes time, weather forecast and attractions nearby.

Figure 11.3 shows the types of context as Human and Environment. Through context-aware sensing and computation, the CPS can acquire contextual information and use it intelligently. These systems will thus be able to anticipate needs and situations of the user and react to the environment around them.

The main aim of the context-aware system is to attain a depiction of the user's view of his/her nearby world. The key issue is to reduce the distance between the user's and the system's view of the real world. The Global Positioning System (GPS) is a deep-rooted system to find the location of the user. There are few sensors that are not clearly understood and interpreted with the sensed information.

The view of a user is based not only on human senses, but also includes the experience and memory of the user. The view of a human has multiple dimensions. In a walk home from the cab stop on a late, dark evening, when the street is silent, one user may feel scared but another user may feel relaxed, calm and cool. It is clear that the availability of the sensor data may not give us a clear idea about the scenario.

Now, it is essential to note that neither system design not its implementation recognizes the environment in the same way as a user. The user's view and the user's knowledge will both contribute to the user's expectations. As explained in the above example, it is difficult to model the perception of every user individually. The system will develop a model of the human as a logical expert system, which may be statistically derived with the common characteristics of a set of users.

The user observes the environment with his/her sensory organs and uses the previous memories and experiences to perceive the context of the current environment. He/she reacts and expects, according to previous events that had happened in the real world. However, the context-aware system utilizes

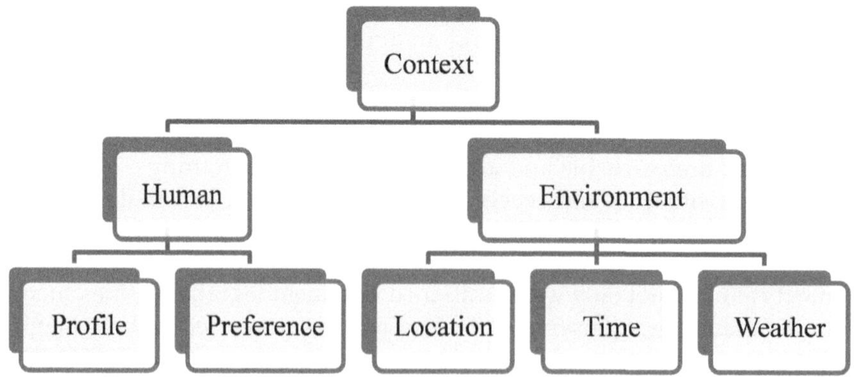

FIGURE 11.3
Types of context

a model of the human which may resemble the common features obtained from a set of people. It obtains the input from various sensors and, along with the human model, will provide the perception of context. Figure 11.4 shows the relationship between the user's expectation of the real world and how that has been modelled in the Context-Aware System.

Currently, there are various types of sensor that are used to obtain the background information. Commonly used sensors are (i) Global Positioning System to sense user location and speed, (ii) microphones to sense sound, (iii) gyroscopes to sense device angular position, movement and vibration and (iv) magnetic field sensors to sense direction. There are also sensors available to monitor the temperature, humidity, proximity and air pressure. The physiological data are collected using Electroencephalogram (EEG) and Electrocardiogram (ECG). To measure the nature of the skin, we can use the Galvanic skin response, that uses the resistance between two electrodes on the skin to find the dryness of the skin, which would be used to identify the feeling or emotion, e.g., surprise or fear. To provide the system with context information, any type of sensor can be used to sense the related information.

To determine "matching" between sensor information and context, we can use machine learning and data mining. By defining a situation, a set of features are described and matched with the sensor input. This is done in the simple rule-based systems. In another way, we can record the conventional situations and calculate delegated features for those situations. The

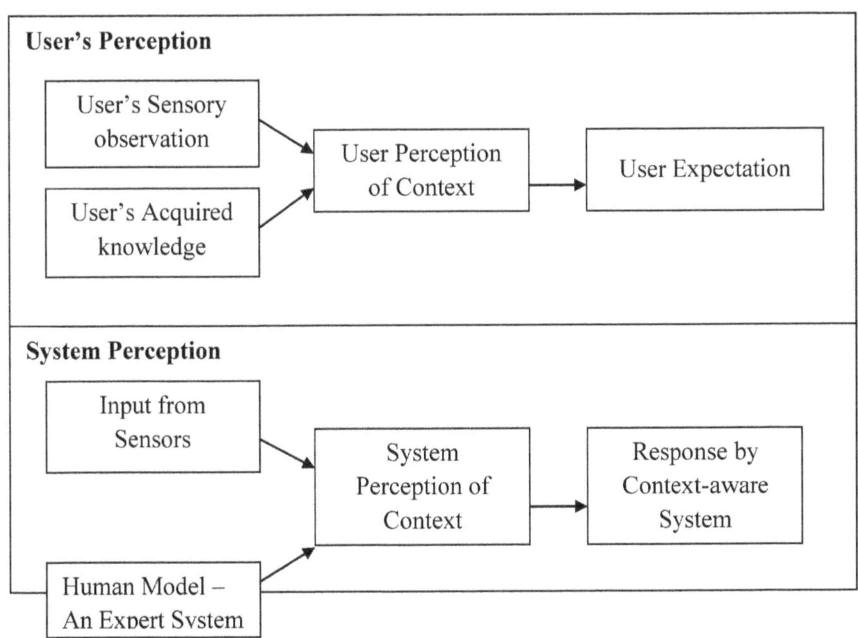

FIGURE 11.4
User perception and system perception

delegated features are calculated and compared with the recorded situations. Then, the current context can be calculated with "nearest-neighbour matching". To determine the system's performance, the performance of the algorithms used to calculate the context should be measured. The classical information retrieval systems can be used as a benchmark to compare the precision and recall of these algorithms. The assessment of the context-aware systems should mainly consider the probability of a particular context, as, otherwise, rare events may be missed.

Table 11.1 shows the context in terms of location, action performed and function mapping carried out by the system for a doctor without disturbing his/her mobility. As he/she enters the hospital, the list of patients waiting for him/her is displayed with their profile. As he/she enters the ward, the details of his/her patients are displayed. When he/she is travelling, his/her location is updated every 60 seconds. When the doctor is away from the hospital, the status of his/her patients admitted to the critical care unit is updated every 60 minutes. Whenever he/she enters the residence the tweets are displayed and whenever he/she exits from the residence the to-do list is displayed. When the doctor enters the gym, the music player is switched ON.

11.4 Generic Sensor API

A sensor is a device which measures the physical quantities and provides a reading that is composed of the value and the time at which the value was read. The sensor types are characterized as low-level sensors or high-level sensors. The sensors that are described by their implementation are called low-level sensors (e.g., gyroscope), whereas those described based on their readings are called high-level sensors (e.g., temperature sensor). There are types of sensors that measure different physical values, such as temperature,

TABLE 11.1

A Mobile Doctor's Context and Function Mapping

Context	Action	Mapped Function
In the hospital	On entry	Show appointment with patients
In the ward	On entry	Show the details of his/her in-patients
In travel	For every minute	The current location of the doctor must be submitted to the Web service
Anywhere	Every 60 minutes	Update the condition of critical patients
At residence	On entry	Show Twitter messages ("tweets")
At residence	On exit	Show to-do list
At the fitness centre	On entry	Play the music player

air pressure, heartrate, motion and so on. The high-level sensors mostly depend on the low-level sensors. For example, to build a pedometer, output of a gyroscope is required.

The sensor fusion is a process of combining the sensor readings from different types of sensors. For example, a patient's blood pressure and heart rate should be sensed to monitor the functioning of heart, and the patient's location should be identified to direct the medical personnel. This is a good example of sensor fusion, which includes different sensors, specifically: (i) a pressure sensor, to sense the blood pressure of the patient, (ii) a heart rate sensor, to determine the number of heartbeats in a specific time period, and (iii) a location sensor, to identify the location of the patient. Depending on the scenario, the sensor fusion can be carried out either at the hardware or software level.

Some sensors have in-built resources that can be used to calibrate and perform sensor fusion at the hardware level. This helps the Central Processing Unit (CPU) to execute in parallel, and reduces the battery power consumption. If it is impractical to perform at the hardware level or if an application-specific fusion method is required, then sensor fusion can be carried out at the software level.

Generic Sensor API is proposed to make both simple and complex use cases. The use case may consider a set of different sensor types, in which case, any one sensor will be defined as a default sensor. If the platform provides the default sensor, then the user needs to accept the specified sensor as the default sensor; otherwise, the user can decide any one sensor to be the default sensor.

```
sensor = new accelerometer ({r: value})
e.g., sensor = new accelerometer ({frequency: 35});
  sensoronreading = 0 => console.log (sensorfrequency)
  sensorstart( );
```

If there are multiple sensors of the same type, then special extensions should be used to specify them separately. For example, if the pressure of the tyre has to be sensed in a car, both the position (front or rear) and the side (left or right) should be specified. It is given as:

```
var sensor = new TyrePressureSensor({position: "rear",
                        side: "left"}.
```

A platform sensor can give the readings to the user agent by taking the "reading change threshold" into account. The reading change threshold is a value that specifies whether the default measures in a sensor are sufficient to update the respective readings. This threshold is based on the software and hardware conditions that prevail in the current situation, as well as the accuracy of the sensor.

11.5 Semantic Sensor Network Ontology

The semantic sensor ontology is the ontology that describes the connected sensors and the data collected by them, their features and also their actuators. SSN (Semantic Sensor Network) follows modularized architecture by including a lightweight but self-contained core ontology called SOSA (Sensor, Observation, Sample, and Actuator) for its elementary classes and properties, as shown in Figure 11.5. Ontology modularization is used to segment ontology into smaller parts. SSN Ontology integrates and maps the available ontologies to create a new ontology. SSN Ontology applies both vertical and horizontal modularization, where vertical means the unidirectional import of one ontology from another, and horizontal means the bidirectional import between any two ontologies. For example, Dolce Ontology imports from SSN Ontology (SSNO) but the reverse does not happen, and hence it is vertical segmentation. The sample relations module depends on SSNO and the reverse also happens, so this is horizontal segmentation.

The sample ontology in Figure 11.6 shows the relationship and connectivity among the objects, such as sensor, observation, stimulus and procedure, followed by the system and the result obtained from the sensor. The stimulus acts as a proxy for the Observed Property, which may be the feature of interest. This sample ontology is referred from www.w3c.org to provide an example for the ontological representation of SSNO.

Several conceptual modules are needed to describe the sensing, actuating and sampling concepts. Those conceptual modules are shown in Figure 11.7, as referred from www.w3c.org.

The system defines the environment in which the sensors and actuators are used to serve a purpose, such as healthcare or weather forecasting. The system is developed only in a procedural way, which defines the input and

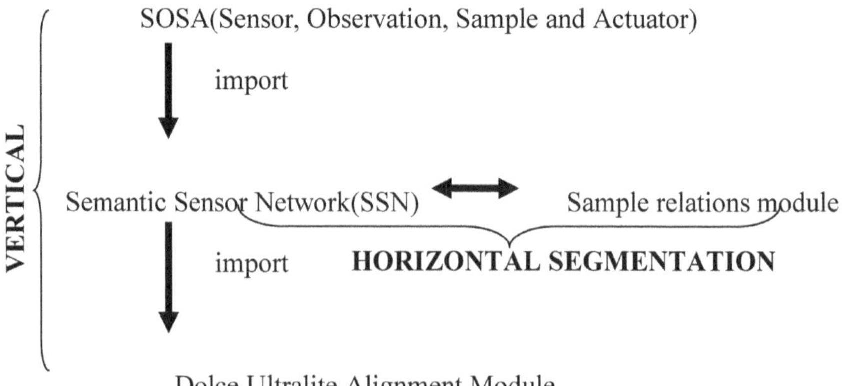

FIGURE 11.5
SOSA and SSN ontologies with their vertical and horizontal modules

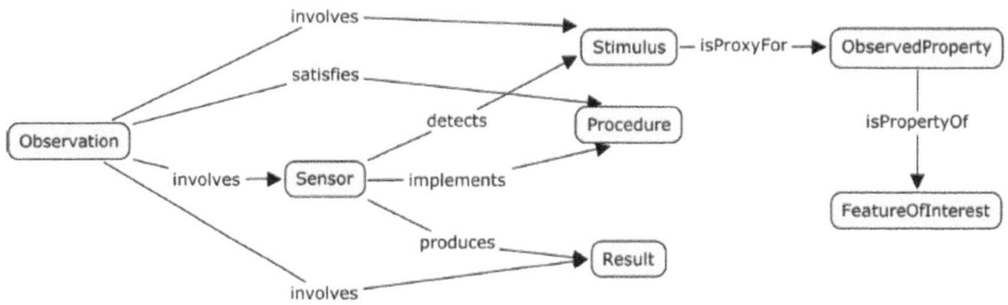

FIGURE 11.6
A sample ontology that shows observations (www.w3c.org)

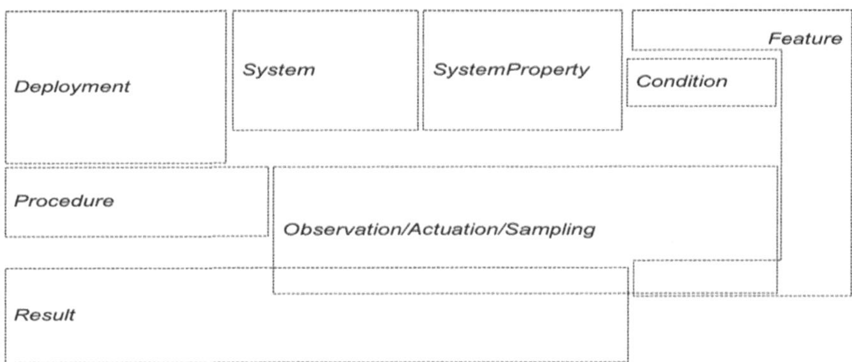

FIGURE 11.7
Conceptual modules for SSN

output of the system. The system's capability is described as the properties of the features, that are relevant and considered in the application. The system contains the sensors that observe the entities, such as date and time, which are described as the observable properties. The system is constrained by the survival range and operating range, which are defined by the condition class. The observation perspective of the overview of SSN classes and their properties are defined in Figure 11.8, as referred from www.w3c.org.

11.6 Healthcare Cyber Physical System (HCPS)

The major changes in the current population are ageing, obesity and seasonal epidemic diseases, which boost the danger of developing various illnesses that need medical intervention. The Government seeks various ways to overcome this with efficient techniques such as "Telemedicine" or "Personal Health System". It is obvious that the proposal explained in this

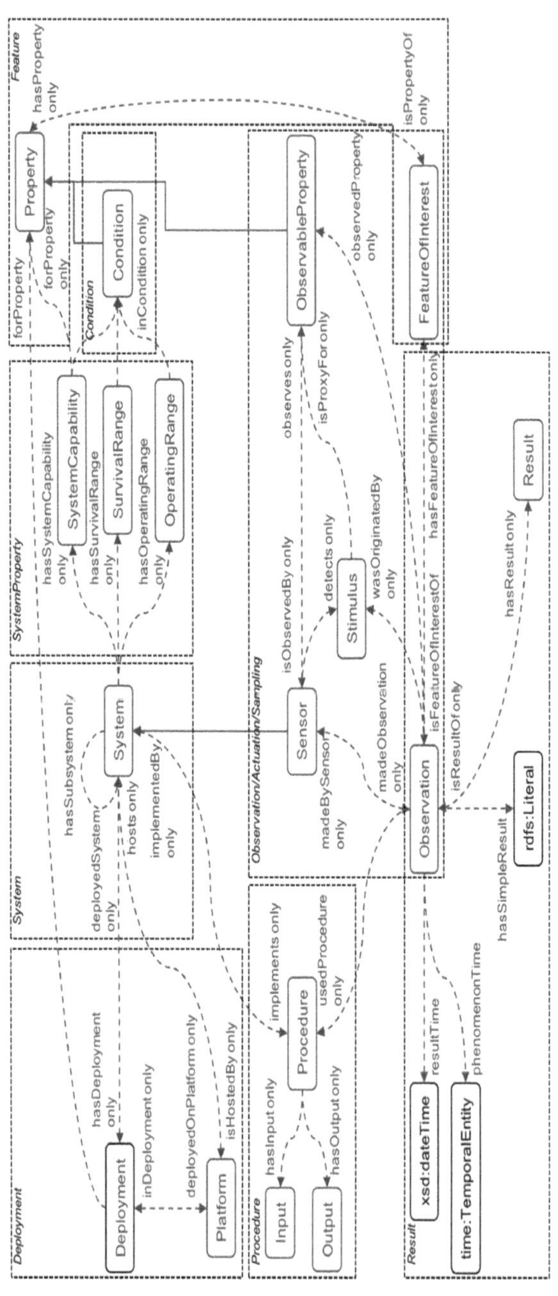

FIGURE 11.8
Overview of SSN classes

section, Healthcare Cyber Physical System (HCPS), would be a new direction of research, which aims to serve the people.

The widespread chronic medical conditions are diabetes, cardiac disease and hypertension, which are twinned together. In order to control these conditions, a continuous monitoring of blood glucose level and blood pressure is essential. The essential sign-observing systems should monitor the following: (i) level of glucose in blood, (ii) pressure level of blood, (iii) pulse rate, (iv) the patterns generated in an electrocardiograph (ECG) device, and (v) the rate and effectiveness of respiration.

There are numerous health issues that need uninterrupted monitoring of the vital signs for diagnosis and relevant treatment. Conventionally, for this, the patient must be hospitalized with costly equipment, and have medical staff readily available. Otherwise, the patient may stay at home but the use of the costly equipment and medical personnel continues in this situation. Recently, much time and effort has been expended to develop small wearable devices, with advantages including reduced cost, greater mobility for the patient and enhanced physiological data for the doctor to investigate the clinical condition of the patient.

To apply Cyber Physical Systems into the healthcare domain, a new method has been proposed and termed the Healthcare Cyber Physical System (HCPS). HCPS is used to monitor and treat the patients by the integration of several medical devices. This system helps to achieve effective, independent and efficient results for both doctors and patients. This may be implemented using the semantic sensor network, the capabilities of which are represented as ontology and exposed to the Web through a Generic Sensor API.

Under critical medical situations, the patient needs continuous monitoring and prompt treatment. In such situations, an intelligent system is required to capture the sensed data and provide referrals for the patient, as guided by the physician. Recently, remote monitoring has become an essential service in the healthcare domain, as it provides care to certain patients, for example, those who are aged but live alone. It is helpful for elders, pregnant women and patients with severe medical conditions, who need nonstop monitoring. The technical advances help to implement HCPS, which assures quick treatment and greater security for patients. But HCPS still has some design issues, as the patient's health depends on systems which need human intervention. In case of emergencies, the control may be lost, affecting the patient severely.

There have been major developments in technology in the field of control systems to substitute for the activities of humans. These developments can make a remarkable contribution in various sectors, including automobiles, construction, banking and healthcare. The control system plays the vital role in Cyber Physical Systems, to sense, compute and communicate with the physical world. Using uninterrupted monitoring and personalized assistance, HCPS aims to provide effective security for the patients. But to provide those abilities, the conventional HCPS system must address the design issues.

Prior to using the system for patients, a risk evaluation must be performed on HCPS to identify any risk factors. The system should satisfy all the essential requirements and also any new requirements which may arise. Usually, such medical devices are approved after an extensive and costly approach. The approval task may be simplified if there is an assurance regarding security concerns. To protect the patient from physical harm, it is essential to protect (i) the privacy of the patient's data collected from various devices to help the care giver and (ii) the medical devices from denial-of-service attacks.

In this chapter, HCPS is augmented by context-awareness, and thus termed Context-Aware Healthcare Cyber Physical Systems (CA-HCPS). The context-awareness is with reference to both the patient and doctor, as the latter is expected to be equipped to handle critical situations with the use of such current technology. Generic Sensor API helps us in gathering the current sensor data and communicating them, to make correct decisions at critical moments of the patient's care. Section 11.7 presents and explains the proposed system architecture, CA-HCPS, in detail.

11.7 System Architecture for Context-Aware HCPS (CA-HCPS)

CA-HCPS is the system proposed for the healthcare domain, by combining context-awareness with Cyber Physical Systems, to provide a dynamic solution for both the patient under critical care and his/her doctor.

Figure 11.9 shows a typical architecture used in Context-Aware HCPS (CA-HCPS), in which sensor and actuator units are tightly coupled with a control unit, and appropriate properties are measured at each system level, so that measure and control loops are performed correctly, using a device such as API/Bluetooth/Zigbee.

In Web Service Enabler, the abilities of the physical entities can be enfolded as services and SOA technologies, such as discovering, selecting and composing those services. The Web service enabler elements are used to create code generators. The Web service is also formed by the enabler through application development.

The Web Service repository have two key types of service: simple and composite. The simple services are again separated into two classes: Web service for interacting with physical devices, and any other services (non-physical Web software services). The physical services are generated by the Web service enabler and registered in the Web service repository.

These Web Services are advertised in the repository. They have to be discovered, based on the functional needs of the consumer (patient/care-giver). The service provider offers a set of services and publishes them in the repository. The requester seeks a service, specifying his/her input and the expected output. A composite service is created by combining elementary

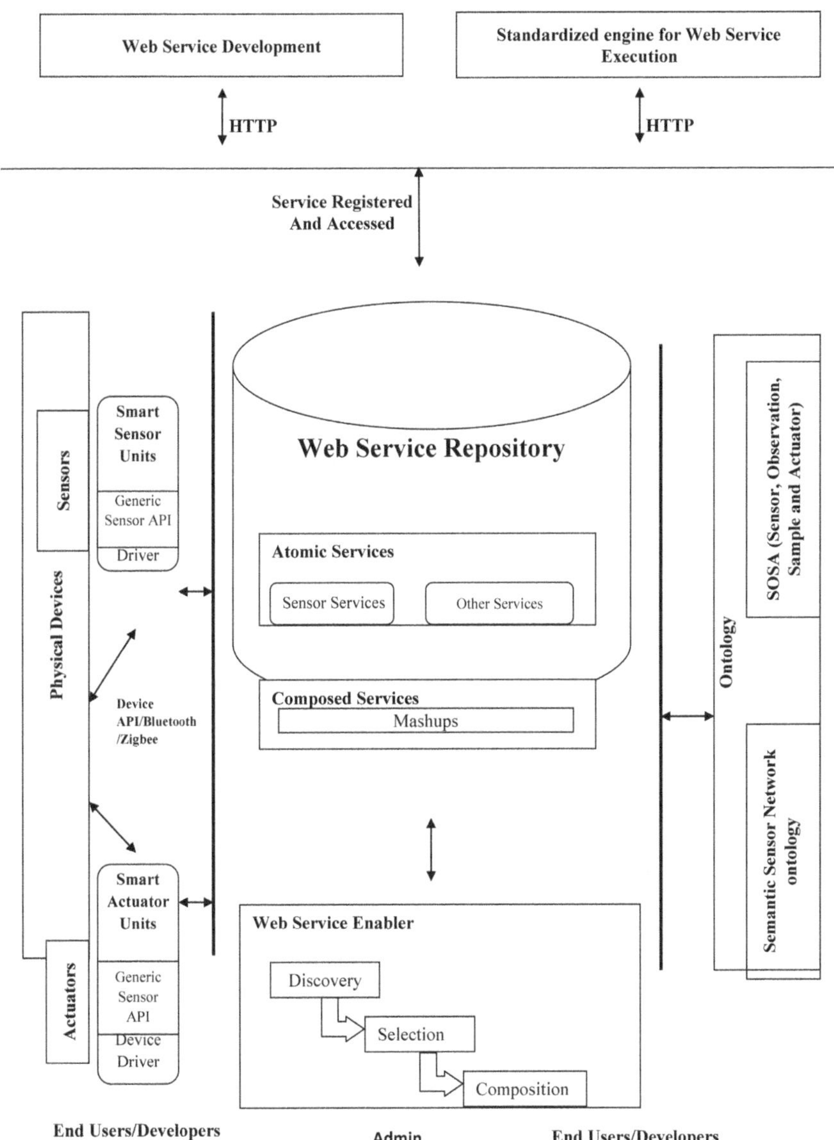

FIGURE 11.9
System architecture for context-aware HCPS (CA-HCPS)

services. A composite service is also considered as it offers easier creation and allocation of applications, along with the end users.

The CPS application developers or the end users can add more composite services to the repository, although the elementary services are managed by the administrator. A standardized engine for Web Service Execution is responsible for carrying out services. It also has an origin and data transmit model that allows various service systems to run at the same time.

11.8 Open Challenges and Opportunities in CA-HCPS

Composability includes the combination of cyber and physical components and the blending of technological systems with the human equivalent. The important challenges are as follows:

- To provide the valuable association between a person and an automated system.
- To confirm the dependability of the proposed system, in terms of privacy.
- To ensure the security from intruders and hackers of devices and software in the system.

The connected medical devices in CA-HCPS give rise to security- and privacy-related issues. An intruder may attack the patient by changing the programme in the devices of the CA-HCPS network. The closed-loop control, automated therapy delivery and alarm capabilities worsen the security issues. Generally, an attacker of CA-HCPS targets one of three choices:

(i) Patient: An attacker directly targets the patient's health by targeting the sensing, processing, communication and treatment delivery aspects of the CA-HCPS. For example, an attacker alters the infusion pump to deliver a larger dosage of medicine than is necessary;
(ii) Data: An intruder retrieves the health data of a patient in an unauthorized manner. This results in loss of privacy with respect to the patient's data, that leads to the risk of patient mistreatment;
(iii) Device: An attacker limits device availability by mounting a denial-of-service attack on the CA-HCPS, so that its performance is downgraded.

The main objective is to find the middle ground for the communication between the CA-HCPS and the organization's internal network to access patient's data and other network operations.

The medical devices, that are networked and interoperable, are capable of being composed dynamically. All these compositions need to be predicted in advance to identify the possibilities of unexpected communication between the devices in the system. One sample scenario says that the networked devices that are attending a patient may interfere with one another as they are close to each other. As a consequence, the treatments can interfere among them and affect the physiological responses of the patient. The designers of CA-HCPS should keep these interferences in mind and make sure that the system providing a treatment is made aware of potentially interfering treatments through adequate contextual information.

The important need of modern medicine is to create medical devices that are able to provide continuous monitoring and care through decision support and proper therapy to the patients. These devices reduce the cost and act as an alternative to home-based or ambulatory care. The patient's health is monitored continuously to provide better treatment at any time. This system should provide event notification, in case of emergencies, and provide the first responders with accurate and complete data. Various diseases, such as cardiac- and neurology-related conditions, are monitored by collecting the state information, such as sleep, awake and fatigue.

In the future, the tools and automation will support humans in Cyber Physical System design to: (i) increase the efficiency and effectiveness of human-to-human communication, (ii) provide accommodation tasks to automate complex tasks, such as exploring the design space, managing changes and verification, and (iii) provide sufficient association between humans and the supporting AI systems. Organizations and management will be tactically prepared by establishing greater awareness of complexity and adopting life-long learning. Awareness relates to creating an understanding of both the benefits and risks associated with complexity.

Creating CA-HCPS requires dealing with various important challenges, which include:

- The medical devices are strongly supported by software. Software is used in place of hardware, especially in terms of safety and security. Efficient software development is critical to ensuring the safety and effectiveness of CA-HCPS.
- Medical devices have interfaces for communication. which should be safe, secure and effective. These devices interoperate with each other to detect diseases and generate alarms during emergencies. It is a difficult task to develop the computationally intelligent software for randomly varying physiological parameters of the human body.
- The increasing autonomy of CA-HCPS, by initiating therapies based on the current state of the patient's health, needs access to medical data. These medical data must be managed with care, as an intruder may access the data and either discriminate, abuse or cause physical damage to the patient. Thus it is essential to maintain the security of CA-HCPS. The dependability of the software used in medical devices is therefore a key issue. The best way to handle this is to obtain certification for medical devices. Obtaining certification, then, is a key issue.

References

1. Dey, Anind K. Understanding and Using Context. *Personal and Ubiquitous Computing*, 5(1). (2001), 4–7. doi:10.1007/s007790170019.
2. Dac Hoang, Dat, Kim, H.Y.P.C.K. Service-Oriented Middleware Architectures for Cyber-Physical Systems. *IJCSNS International Journal of Computer Science and Network Security*, 12(1). (2012), 79–87.
3. Gunes Volkan, Peter, Steffen, Givargis, Tony, Vahid, Frank. A Survey on Concepts, Applications, and Challenges in Cyber-Physical Systems. *KSII Transactions on Internet and Information Systems*, 8(12). (2014), 4242–4268. doi:10.3837/tiis.2014.12.001
4. Huertas Celdrán, Alberto, Manuel Gil, Pérez, Félix J. García Clemente, and Gregorio Martínez, Pérez. Sustainable Securing of Medical Cyber-Physical Systems for the Healthcare of the Future. *Sustainable Computing: Informatics and Systems*, 19. (2018), 138–146. doi:10.1016/j.suscom.2018.02.010.
5. Kim, Kyoung-Dae & Kumar, Panganamala. Cyber–Physical Systems: A Perspective at the Centennial. *Proceedings of the IEEE - PIEEE*, 100(Special Centennial Issue). (2012), 1287–1308. doi:10.1109/JPROC.2012.2189792.
6. Lee, Edward A. Cyber Physical Systems: Design Challenges. In: *Proceedings of the 2008 11th IEEE Symposium on Object Oriented Real-Time Distributed Computing (ISORC '08)*. IEEE Computer Society: Washington, DC, USA, 2008, 363–369. doi:10.1109/ISORC.2008.25.
7. Schmidt, A., Beigl, M., Gellersen, H. There Is More to Context than Location. *Computers and Graphics*, 23(6). (1999), 893–901.
8. Schilit, Bill, Adams, Norman, Roy, Want. Context-Aware Computing Applications. In: *Proceedings of The IEEE Workshop on Mobile Computing Systems and Applications*, 1995, 85–90. doi:10.1109/WMCSA.1994.16.
9. http://www.w3.org/TR/2004/REC-owl-guide-20040210/
10. https://www.w3.org/TR/2019/CR-generic-sensor-20191212/
11. https://www.w3.org/TR/2017/REC-vocab-ssn-20171019/

12

Intelligent Social Networking in CPS

S. Hemkiran and G. Sudha Sadasivam

CONTENTS

Organization of the Chapter

Section 1 provides the terms and terminologies, to help the reader to understand this chapter. **Section 2** presents an introduction to cyber physical social systems, highlighting the applications, challenges and opportunities that exist in this realm. **Section 3** outlines IoT in the context of social networks. **Section 4** discusses community detection, with the emphasis on smart communities. **Section 5** presents the link prediction models utilized in CPS and provides a brief overview of the different metrics. Finally, **Section 6** presents the conclusion of this chapter.

12.1 Terms and Terminologies

Cyber Physical Systems (CPS): A confluence of cyber systems encompassing computers and communication devices together with physical systems consisting of sensors, actuators and users.

Social networking: Using Internet-based social media sites to stay connected with friends, family, colleagues, customers, or clients.

Smart community: A community consisting of government, business, and, residents who decide to use information technology to transform life /work in their region.

Sensors: Devices that measures a physical property and responds to it.

Social computing: Aims to recreate social conventions and social contexts using software and technology.

Community Detection: Identifies highly connected groups of individuals or objects inside a network.

12.2 Introduction

Cyber physical systems (CPS) are a confluence of cyber systems, encompassing computers and communication devices, together with physical systems consisting of sensors, actuators and users. The interactions between these systems are tightly integrated at multiple scales and levels by a computing and communicating core. The operations of CPS are monitored, controlled and co-ordinated in realtime, using computer-based algorithms securely linked through the Internet (Ashibani et al. 2017; Lee et al. 2015; Rajkumar et al. 2010), and consist of a feedback loop, as shown in Figure 12.1. In CPS, the sustainability, safety and efficiency of interactions between the physical and cyber world is achieved by actuation of unconventional objects, such as machines and switches, using the Internet (Negenborn et al. 2014; Sisinni et al. 2018). The term CPS initially originated in the USA in the year 2006.

As shown in Figure 12.2, CPS connects the cyber and physical world through three key enablers, namely, computation, communication and control (Gao et al. 2013). This is widely referred to as the 3C architecture of CPS (Liu et al. 2017; Xiong et al. 2015). The development of the CPS domain requires an integration of expertise across engineering disciplines, such as human–computer interaction, learning theories, material science and biomedical engineering. Traditionally, physical processes and control flows are modelled using differential equations and automata, respectively. Whereas such modelling approaches may suffice for component level associations, these approaches exhibit shortcomings during system-level physical and behavioral interactions between distinct components in CPS (Baheti et al. 2011).

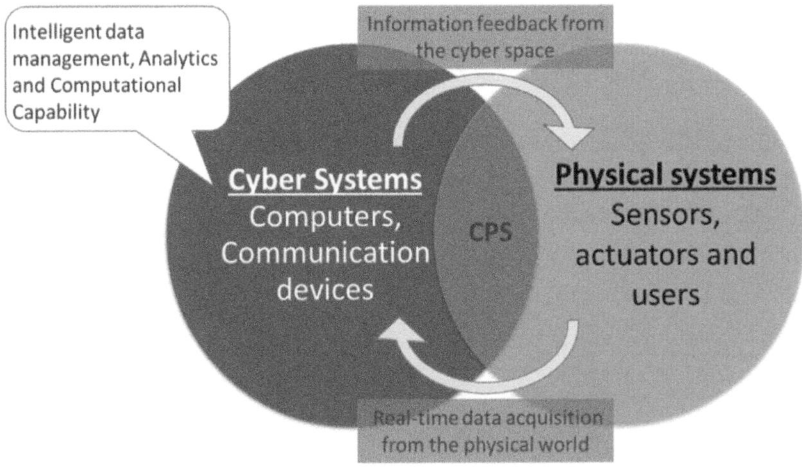

FIGURE 12.1
Components of CPS

In spite of being an emerging topic, the system components of CPS are well defined, as depicted in Figure 12.3. The physical world, interfaces and cyber systems constitute the components of CPS. The devices and entities that will be monitored or controlled are designated as the physical world. Embedded devices, incorporating cutting-edge technology which can exchange data with their distributed environment, are termed cyber systems. The intermediate components, such as the sensors and actuators, analogue-to-digital (A/D) and digital-to-analogue converters (D/A), which aid the cyber systems to communicate with the physical world, are referred to as interfaces. Sensors and actuators are widely utilized in CPS to reciprocally convert from a multitude of energy forms to analogue signals in the form of electricity (Gunes et al. 2014).

The forthcoming sections present the need for CPS and its various characteristics.

12.2.1 Need for CPS

The advent of the World Wide Web in the 1980s–1990s led to an exponential increase in the usage of computers for day-to-day applications. Simultaneously, advances in hardware, such as shrinking memory spaces, better graphical interfaces, affordable sensors and higher computational speeds, resulted in extensive, innovative applications, such as digital libraries, online communities, e-commerce and social networks (Rajkumar et al. 2010). These advances were aided by the rapid growth in technology, such as sources of renewable energy, augmented internet bandwidth, portable laptops and smartphones. The proliferation of the Internet, coupled with technological advances, have resulted in the demand for linking the cyber and

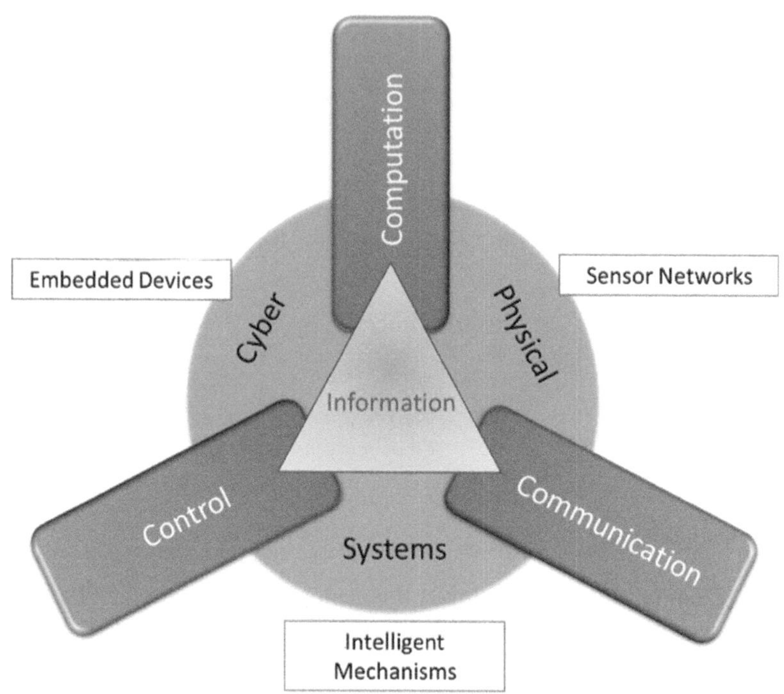

FIGURE 12.2
3CArchitecture of CPS

physical world in domains such as aerospace, transportation vehicles, industrial automation, defence systems, safety-critical processes, medical devices and healthcare. CPS can therefore be envisioned as embedded systems with real-time high-power computing capabilities, blended with distributed sensors and controls. CPS have wide-ranging applications, from compact heart pacemakers to expansive power grids.

12.2.2 Characteristics of CPS

In order to perform functions, such as collecting data, monitoring and controlling systems, a CPS should pass with respect to specific issues, namely flexibility, security, usability and privacy concerns (Xiong et al. 2015). Consequently, a system is required to possess the following characteristics (Liu et al. 2017) to be categorized as a CPS, as shown in Figure 12.4.

12.2.2.1 Physical System as the Principal Component

Physical systems consist of perceptive devices with sensing, computing and wireless communication capabilities, such as sensors, Global Positioning System (GPS), Radio Frequency Identification (RFID) tags and actuators.

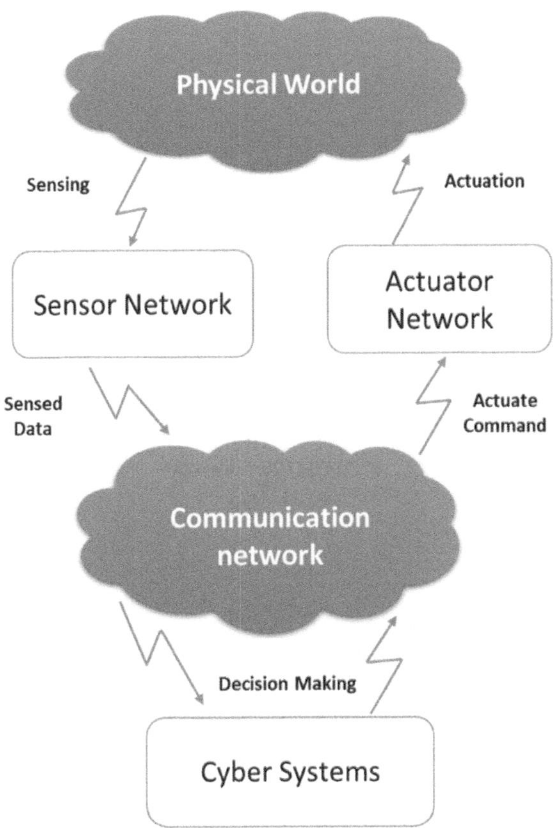

FIGURE 12.3
Holistic view of CPS

The physical system is the principal component of any CPS, as it is responsible for recognizing and interpreting real-time data collected from the physical world (Ashibani et al. 2017). The data, gathered by different sensors in the form of sound, vibration, temperature, location, humidity etc., is then transmitted *via* communication networks such as Infrared, 4G internet, Bluetooth and WiFi, in order to be assimilated and analyzed by the information system. Due to the wireless nature of these networks, reliability of connections and, therefore, predictability of interactions is an issue in CPS. Particularly, the use of mobile devices leads to great uncertainties in the network. For instance, connected automobiles in a vehicular network may interact between one another only when they are at close range.

12.2.2.2 Integration with Information System

The information system can be envisioned as the central hub of any CPS, as this system should be competent to handle the large volume of data

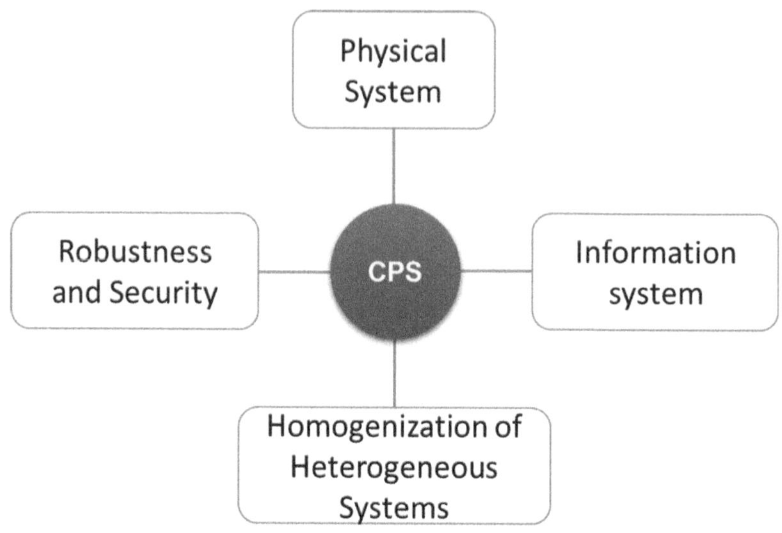

FIGURE 12.4
Characteristics of CPS

generated by a multitude of physical systems, and simultaneously provide real-time feedback. Additionally, information systems should possess the ability to handle the various tradeoffs, such as network latency, memory management, data validation and reconciliation (DVR) and hierarchical storage management (HSM). A close integration between the physical and information modules is achieved, using embedded systems.

12.2.2.3 Homogenization of Heterogeneous Systems

A CPS is not a stand-alone system and is an intelligent network of heterogeneous distributed systems, such as sensors, high-power servers and handheld mobile devices. In a large-scale CPS, with components distributed world-wide, spatiality and time synchronization are of paramount interest during integration and interaction among heterogeneous systems.

12.2.2.4 Robustness and Security

In contrast to online systems integrated over the internet, CPS encompasses a host of hardware, users and software systems, communicating with each other. This necessitates infallible integration at a deeper level. Therefore, robustness and security are key requisites of CPS. Robustness is the ability of a processing system to deal with errors during input and execution. In a network, data authentication, validation and verification of the sending device's identity are the principal factors to be examined during data input. The receiving device must decipher the true identity of the sender to avert bogus data being transferred and executed. Hence, security of the information transmitted over the network should be ensured, using appropriate encryption and decryption algorithms. Another aspect of a robust system is the ability to predict requirements and to scale resource allocations to fulfil real-time tasks. In essence, a large-scale CPS should possess the ability to dynamically adapt, reconfigure and reorganize, in addition to manifest attributes, such as robustness, security, reliability, interoperability and credibility.

CPS have been extensively deployed in cross-domain applications involving autonomy, interoperability and adaptability, such as entertainment services, industrial mass production and intelligent transportation systems. Furthermore, CPS can be envisioned as the elementary form of the Internet of Things (IoT) since, in CPS, real-world objects are mapped to the cyber world as cyber entities (Wang et al. 2015; Ning et al. 2016).

12.2.3 Introduction to Cyber Physical Social Systems (CPSS)

As mentioned earlier, CPS is an indivisible integration of cyber and physical systems. These systems are designed and built to interact with humans. The contemporary surge in the utilization of social networks generates an abundance of user-specific data, which can be integrated with CPS to provide large-scale, real-time services, such as air pollution monitoring, venue recommendations, smart parking systems, smart cities and smart homes. This paradigm of taking into account the social dynamics as a part of CPS is referred to as Cyber Physical Social Systems (CPSS)(Amin et al. 2019). CPSS can be perceived as a three-dimensional space, incorporating close interaction and involvement of humans with the physical and cyber space. CPSS act as the fundamental enabler of social computing. In CPSS, the physical data shared by users in a network is collected using numerous sensors. In conditions, such as commenting and recording preferences, the users themselves act as sensors (Zheng et al. 2017). Social computing typically consists of two aspects, namely dynamic social interaction and sensing social phenomena. Whereas social interaction pertains to interpreting data gathered from immense online interactions among users, sensing social phenomena refers to analyzing and comprehending the pattern of interaction among users (Zeng et al. 2016). A CPSS, while amalgamating the four realms of information, physical, social and cognitive domains, also performs auto-synchronization and parallel processing. Therefore, CPSS are utilized in applications involving real-time command and control, such as military operations (Liu et al. 2011). In addition to the physical entities in CPS, social attributes, such as affiliations and ownerships, are appended, leading to an interfusion of CPS and social realms to form CPSS, an enhanced version of IoT (Ning et al. 2016).

CPSS consists of a deeper integration and semantic interdependence between physical layers, composed of sensors, and collective intelligence, signified by futuristic cognition to blend machine and human perceptions (Sheth et al. 2013). A major concern when supplementing the existing CPS with social data, is the breach of privacy of users. When blending physical and social systems, confidential data such as individual habits, user location, stay and travel details, are evidently shared through the cyber world. Unlike privacy in social networks, long-term physical profiles and behavioral patterns of users are expected to be privacy preserved in CPSS. Handling of the enormous data thus generated, is one major challenge in CPSS (Zheng et al. 2017).

12.2.3.1 Applications of CPSS

CPSS have been beneficially applied to numerous fields. Table 12.1 summarizes the field and scale of application in various sectors.

12.2.4 Challenges and Opportunities in CPSS

Several challenges are yet to be addressed for large-scale implementation of CPSS to solve day-to-day problems. Table 12.2 summarizes a few of the problems outlined in the literature.

12.3 Social Internet of Things (SIoT)

IoT is an interconnected structure of computing devices, humans and digital objects, with the ability to communicate over the Internet. Conventionally, the term 'things' refer to interactions in cyber space, cognitive reasoning,

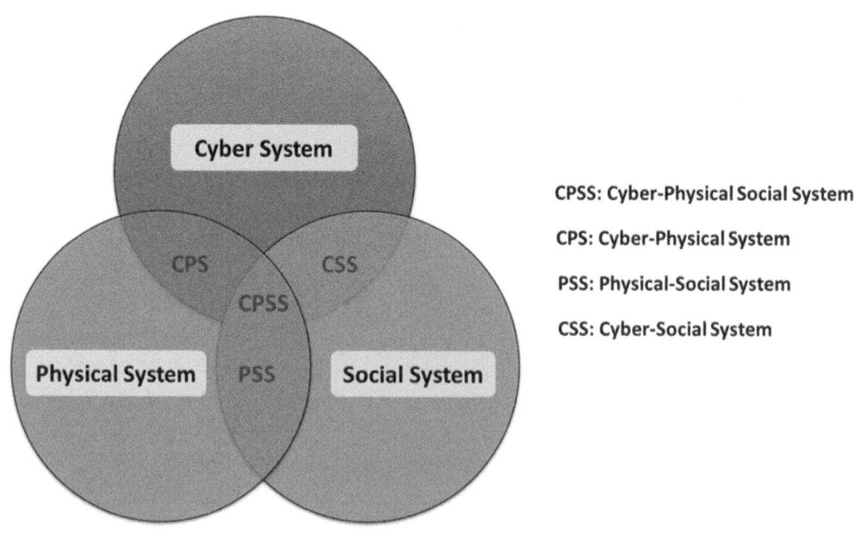

FIGURE 12.5
Scope of CPSS

TABLE 12.1

Applications of CPSS

S. No.	Field of Application	Scale of Application: Small	Medium	High	Function	References
1.	Healthcare		✓		• Monitor health condition of the patients • Suggest proper treatment through smart medical devices.	Sheth, Anantharamand Henson (2013), Gunes et al. (2014)
2.	Intelligent transportation		✓	✓	• Improve safety and traffic management skills in real time.	Elmaghraby and Losavio (2014), Han, Duan and Li (2017)
3.	Robot utilization	✓	✓		• Semi-automatic or fully automatic type utilized for human welfare.	Gunes et al. (2014), Ning et al. (2016)
4.	Vehicular networking systems		✓	✓	• To monitor the conditions of roads and their issues.	Xiong et al. (2015), Meenalaxmi, Abinaya and Suseela (2018)
5.	Education based on mobile			✓	• Affordable e-learning as an effective way of acquiring knowledge	Ning et al. (2016), Guo et al. (2017)
6.	Defence Systems	✓			• To monitor military aviation systems and their traffic regularities	Wang, Törngren and Onori (2015), Han, Duan and Li (2017)
7.	Complex infrastructure – smart grids	✓	✓		• To contribute to the welfare of treasured properties and safety of nation	(Wang, Törngren and Onori (2015), Liu et al. (2017)
8.	Construction automation	✓			• Utilized for next generation building automation to control heating, cooling and electricity consumption	Rajkumar et al. (2010), Baig et al. (2017)
9.	Smart homes		✓	✓	• Utilized to enhance safety and comfort of home and society at large	Li et al. (2011), Elmaghraby and Losavio (2014)

TABLE 12.2
Challenges in CPSS

S. No.	Challenges	Problem Delineation	References
1.	Security	Attack happens during transmission of data.	Guo et al. (2017)
2.	Energy consumption	Higher power is consumed by the cyber components during operation.	Shakshuki, Malik and Sheltami (2014)
3.	Dynamic environment	Errors in input or during execution lead to robustness issues.	Wang, Törngren and Onori (2015), Sisinni et al. (2018)
4.	Stability	Problem arises when an operating system crashes, where there are networking issues	Meenalaxmi, Abinayaand Suseela (2018), Guo et al. (2017)
5.	Dependability	Timing inaccuracies for capturing sensor readings	Gunes et al. (2014), Ashibani and Mahmoud (2017)
6.	Maintainability	Certain components paves way for continuous repairing	Gunes et al. (2014), Lee, Bagheri and Kao (2015)
7.	Availability	Certain attacks that may occur in cyber components leads to the non-availability of the system when required.	Ashibani and Mahmoud (2017), Lu (2012)
8.	Safety	May harm when it is operated due to the lack of precise monitoring of the system, affecting the processing in reality	Rajkumar et al. (2010), Sisinni et al. (2018)
9.	Reliability	As CPS includes many components to perform both in the physical and cyber aspect, the control flow between the components may vary, leading to a reduction in reliability factor.	Xiong et al. (2015)
10.	Robustness	Unpredictable environmental occurrences	Gao et al. (2013), Ashibani and Mahmoud (2017)

physical perceptions and social relations. The interactions between the components in this system can be remotely monitored and controlled. They also possess a high degree of independence from human intervention. Furthermore, an interfusion of the physical environment and the virtual world of information can be achieved, utilizing IoT (Yang et al. 2018). The term IoT was conceived by Kevin Ashton in 1999.

IoT facilitates the wireless communication of physical objects, such as sensors, gadgets, automobiles and buildings, with the cyber world, consisting of computers and networks with minimal human intervention. IoT wirelessly connects smart devices, involving sensors, to the cyber world, using Internet (Li et al. 2011; Baig et al. 2017). This interlink, while being beneficial in many aspects, leads to security and privacy issues (Harel et al. 2017). Security solely permits the authorized users to decrypt confidential information and ensures integrity of service, whereas privacy refers to the share-ability of personal information, such as photographs, activities and locations with third-party service providers (Elmaghraby et al. 2014). A deficit in security or privacy leads to vulnerabilities in the system and large-scale breaches in data confidentiality of service users.

In this digital age, social networks are a ubiquitous mode of communication. In addition to publicizing information by virtually connecting people, social networks help to shape and sway user opinions. When conjoined with IoT, the Social Internet of Things (SIoT) offers a multitude of benefits, such as instantaneous decision making and enhanced data access at economical processing costs. SIoT can be envisaged as a unique subset of IoT, wherein numerous interconnected, heterogeneous IoT devices socialize and cooperate with each other in order to accomplish specific tasks in a collaborative manner. The salient feature of SIoT is the formation of social relationships between the heterogeneous IoT devices to provide autonomous services. Such relationships are achieved by means of a service-oriented architecture and lead to a multitude of IoT-powered applications, namely smartgrids, communities and cities (Wang, et al. 2015). In particular, International Business Machines (IBM) relates smart cities consisting of humans and various components, with being intelligent and interconnected with instrumentation (Elmaghraby et al. 2014). In smart communities, IoT sensors are deployed effectively to monitor and respond to changes in their functional environment in realtime (Baig et al. 2017).

Cyber Physical Social Systems (CPSS) consider the virtual locations of the users in addition to their physical locations. Hence, CPSS can be regarded as the third evolution from online social networks. The first generation of broadly utilized social networks, namely LinkedIn, Facebook and Orkut, forged connections between users solely based on actual, existing relationships, such as family, friends and existing business relationships. New connections were created through explicit requests from users. A review of the literature indicated that less than 10% of connections were established between users hitherto unfamiliar with each other. With sustained innovation, the second generation of social networks, also referred to as location-based social networks (LBSN), considered the geolocations of the users. In addition to explicit requests, new contacts could be established between users based on shared locations. Examples of such networks include Facebook Places, Skout and Foursquare. Evolution then led to the third generation of social networks, which utilized the physical and virtual context of the users. Whereas geolocations of the users contribute to the physical context, the data gathered from user behaviour, such as current interests, history of information searched online, liked content and frequently viewed videos or webpages visited, contribute to the virtual context. Due to this interaction between the cyber and physical world, the third generation is often referred to as CPSS (Weth et al. 2017).

12.4 Community Detection

A community is generally defined as a social unit of entities having shared, common thoughts or identities. The size of the community is contingent on the number of members sharing similar interests.

12.4.1 Need for Communities to be Detected

In a social network, community detection is an essential procedure to detect individual nodes that overlap on the basis of functional subunits. Links between communities in biological networks, comprising metabolic networks and protein–protein interactions, demonstrate that large social networks consist of community structures that are organized in a hierarchical manner (Ahn et al. 2010).

12.4.2 Smart Communities

Smart communities refer to the interdependence between humans and physical entities. In particular, 'smart' refers to the ability of a system to acquire knowledge about its surroundings and to employ this vital information to enhance the quality of life of its inhabitants. This interaction can be achieved through an amalgamation of social computing techniques with CPS. A smart community can also be visualized as a multi-hop network of wirelessly interconnected homes (Li et al. 2011). Social networks are typically built on a multitude of relationships, such as business associations, peer groups, friends and familial connections. Several social networking applications have harnessed the aforementioned relationships to create virtual communities. Such communities enable information sharing and widespread dissemination among its members (users). The use of this information by computers, networked *via* the Internet to interact with the physical world by means of sensors, monitors and controls, is the genesis of smart communities (Xia et al. 2011). In essence, smart communities can be envisaged as social objects, such as humans and physical entities communicating with each other and delivering multifaceted services, utilizing social intelligence and a cyberphysical network. In a community, data on user behaviour is gathered from multiple sources. A user can therefore be visualized as a logic unit with attributes, such as location in the physical space, roles assumed in the community and digital information published in the cyberspace. Assimilation of data from discrete users in a community is complex and requires analysis on multiple variables and dimensions. This necessitates link prediction techniques to utilize community-based classification of discrete user activities in order to rationalize analysis of big data from large community-based networks (Zhang et al. 2014).

Numerous techniques, such as pattern mining, clustering and graph-based approaches, have been utilized for analyzing communities and detecting smart, cohesive clusters or subgroups that are closely interconnected in a network. In the pattern mining technique, a community evaluation function is optimized to identify densely connected nodes (Atzmueller 2014). Predictions based on extensive information gathered from community activity are significantly valuable owing to the reasons mentioned hereunder.

- Reduced complexity of analysis: A single model can be created based on big data gathered from communities, comprising numerous individuals, grouped by common behavioural traits. Analyzing this model is uncomplicated, compared with interpreting data of all individuals in a network.
- Trade-off of efficiency *vs* accuracy: Community-based models can expeditiously discover and predict the demands of a majority of the population in a community. However, the accuracy of such models is a contentious issue. In essence, a study of the population requires a more efficient prediction, rather than an accurate prediction, of community activities (Zhang et al. 2014).

A smart city is a class of a smart community, utilizing data and insights from IoT systems, to provide enhanced services to its inhabitants. The framework for a smart city comprises of a network of sensors and actuators spread over the city terrain, constantly communicating with mobile devices *via* cloud-based Internet services. The information disseminated through this cyber physical system assists in monitoring and controlling a city's resources and services, such as smart parking and street lighting, air–water quality, surveillance and product delivery using unmanned aerial vehicles, structural health monitoring, real-time traffic monitoring and smart electricity grids (Cassandras 2016). The U.S. National Institute of Standards and Technology (NIST) specifies a globally accepted standard, comprising six categories of smart city functions, namely environment, mobility, economy, governance, people and living (Baig et al. 2017). A smart environment can be defined as one which acquires and utilizes the knowledge of the surroundings to enhance the environmental circumstances of its inhabitants (Cook et al. 2007).

12.5 Link Prediction

A social network is a realm, comprising perpetual interactions among its constituents. Conventionally, in a social network, the relationship between people is depicted by nodes and edges. Nodes, or vertices, represent the persons involved in the interaction. Edges, or links, denote the relationship that is established between them. Figure 12.6 shows a typical representation of a social network, highlighting the nodes and edges.

The interactions among the multitude of constituent nodes in a social network leads to the formation of communities with shared interests. In order to form communities or groups, it is essential to identify entities with similar interests and to predict formation of interconnections between them. Therefore, knowledge of link prediction is of vital importance for social network analysis. Link prediction denotes the probability of adding new links to an already-existing network at a future time. By identifying the specific connections (links) that will be established in the near future, formation of new relationships can be determined and predicted in the community structure.

Numerous models for predicting the occurrence of links in CPS have been reported in the literature. A Bayesian network model was utilized to create a dynamic traffic surveillance network for an intelligent traffic system (Geng et al. 2018). The method explored real-time prediction and analysis of complex events. The results demonstrated that Bayesian network models were effective in predicting events in CPS. An integrated collaborative filtering recommendation (ICFR) approach was explored for large dynamic networks, such as CPS (Xu et al.2017). This approach generated valuable information from big data to create recommendations, based on similarity between users and

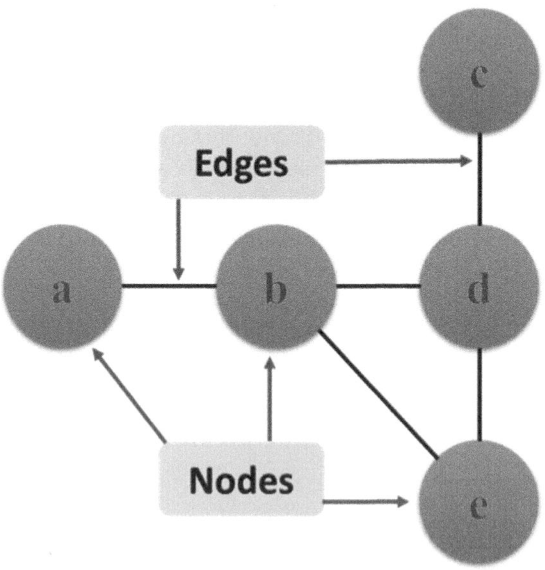

FIGURE 12.6
Representation of a social network

items. The study suggested that ICFR reduced mean absolute error (MAE) by approximately 27.5 % and root mean square error (RMSE) by nearly 15.7 %, suggesting the merit of this approach for CPS. Furthermore, utilizing data from the Twitter network, a geosocial analysis model was investigated in order to study the propagation of information and evolution over time in the cyber and physical spaces (Croitoru et al. 2014). The study highlighted the role of hybrid networks and the intricate nature of interactions in CPS. The performance of the various models utilized in CPS is evaluated using metrics. Table 12.3 lists the prominent metrics utilized in link prediction, as outlined in the literature.

12.5.1 Metrics Used for Predicting Links

The commonly reported metrics for predicting new links are Common neighbours (CN), Jaccard coefficient (JC), Adamic/Adar (AA), Preferential attachment (PA), Resource allocation and Random walk with Restart. In a social networking structure, all the nodes (people) may not be connected together. The CN metric works on the concept of similar interests exhibited by diverse nodes. If common interests, such as similar hobbies exist, a higher probability of group formation exists. Additionally, if two people, 'A' and 'B', have several friends in common, the chance of link formation between these people in the future is higher. JC is a statistical measure that is used for computing the similar features of the sample sets. Whereas the metric AA assigns more weight to the rarer features, the metric PA follows the phenomenon of "the rich get richer". This phenomenon signifies that users with many friends tend to create more connections in the future. The metric RA has been reported to outperform all other metrics when predicting links in the human contact network. The Random walk-based method is preferred to calculate the link relevance in geo-friend recommendation frameworks. Of the metrics outlined above, all metrics, except PA, perform almost identically when detecting similarity in online social networks.

12.6 Conclusion

Cyber Physical Systems embody an integral combination of network and physical systems. However, it is arduous to construct a robust, error-free system with the characteristics of distributed, real-time features by combining physical and cyber resources. As discussed earlier, the varied social networks with which a CPS interacts presents challenges ranging from dynamically predicting link formation to privacy and security issues. Furthermore, due to the multiple components involved, CPS is incapable of exhibiting similar computational processing as would be the case of a traditional system.

TABLE 12.3

Prominent Metrics Utilized in Link Prediction

Link Prediction Metrics	Score (x, y)	Type of Network	Purpose	References
Common neighbours	$\mid \Gamma(x) \cap \Gamma(y) \mid$	Drug target network	Produces the precision level of drug-target interactions much better than the Restricted Boltzmann machines (RBM)	Socievole, De Rango and Marano (2013), Lu, Guo and Korhonen (2017)
Jaccard coefficient	$\frac{\Gamma(x) \cap \Gamma(y)}{\Gamma(x) \cup \Gamma(y)}$	Friends recommendation framework in biology field	For retrieving the similarity between the item sets of users (item sets means the list of preferences by the user).	(Xie (2010)
Adamic/Adar	$\sum_{Z \in \Gamma(x) \cap \Gamma(y)} \frac{1}{\log(\Gamma(Z))}$	Biological network based on protein–protein interaction dataset	Used for predicting future links	(Kashima and Abe (2006), Socievole, De Rango and Marano (2013)
Preferential attachment	$\mid \Gamma(x) \mid * \mid \Gamma(y) \mid$	Facebook dataset	Recommending the future friendship between various users	Socievole, De Rango and Marano (2013), Gupta, Pandey and Shukla (2015)
Resource allocation	$\sum_{Z \in \Gamma(x) \cap \Gamma(y)} \frac{1}{\Gamma(Z)}$	Human contact networks	For identifying the significance of online social similarity	Socievole, De Rango and Marano (2013)
Random walk with Restart	$q_{ab} + q_{ba}$ where q_{ab} is the probability of moving random from node a to b.	Geo-friend recommendation framework (GEFR)	Used for calculating the link relevance between different nodes and providing the friend recommendation related with the user's query	Yu et al. (2011)

Additionally, it is laborious to deal with heterogeneous types of resources applied in a dynamic environment. Nevertheless, CPS have the potential for numerous real-world applications and exhibit a host of benefits. The aforementioned challenges need to be overcome in order to apply CPS on a large scale.

References

Ahn, Yong Yeol, James P. Bagrow, and Sune Lehmann. 2010. "Link Communities Reveal Multiscale Complexity in Networks." *Nature* 466(7307): 761–764. doi:10.1038/nature09182.

Amin, Farhan, Awais Ahmad, and Gyu Sang Choi. 2019. "Towards Trust and Friendliness Approaches in the Social Internet of Things." *Applied Sciences* 9(1): 166. doi:10.3390/app9010166.

Ashibani, Yosef, and Qusay H. Mahmoud. 2017. "Cyber Physical Systems Security: Analysis, Challenges and Solutions." *Computers and Security* 68: 81–97. doi:10.1016/j.cose.2017.04.005.

Atzmueller, Martin. 2014. "Social Behavior in Mobile Social Networks: Characterizing Links, Roles, and Communities." In: *Mobile Social Networking: An Innovative Approach*, edited by A. Chin and D. Zhang, 65–78. New York: Springer Science Business Media. doi:10.1007/978-1-4614-8579-7.

Baheti, Radhakisan, and Helen Gill. 2011. "Cyber-Physical Systems." *The Impact of Control Technology, IEEE* 12(1): 161–166.

Baig, Zubair A., Patryk Szewczyk, Craig Valli, Priya Rabadia, Peter Hannay, Maxim Chernyshev, Mike Johnstone, et al. 2017. "Future Challenges for Smart Cities: Cyber-Security and Digital Forensics." *Digital Investigation* 22: 3–13. doi:10.1016/j.diin.2017.06.015.

Cassandras, Christos G. 2016. "Smart Cities as Cyber-Physical Social Systems." *Engineering* 2(2): 156–158. doi:10.1016/J.ENG.2016.02.012.

Cook, Diane J., and Sajal K. Das. 2007. "How Smart Are Our Environments? An Updated Look at the State of the Art." *Pervasive and Mobile Computing* 3(2): 53–73. doi:10.1016/j.pmcj.2006.12.001.

Croitoru, Arie, N. Wayant, A. Crooks, J. Radzikowski, and A. Stefanidis. 2014. "Computers, Environment and Urban Systems Linking Cyber and Physical Spaces through Community Detection and Clustering in Social Media Feeds." *Computers, Environment and Urban Systems.* doi:10.1016/j.compenvurbsys.2014.11.002.

Elmaghraby, Adel S., and Michael M. Losavio. 2014. "Cyber Security Challenges in Smart Cities: Safety, Security and Privacy." *Journal of Advanced Research* 5(4): 491–497. doi:10.1016/j.jare.2014.02.006.

Gao, Sheng, Hao Luo, Da Chen, Shantao Li, Patrick Gallinari, Zhanyu Ma, and Jun Guo. 2013. "A Cross-Domain Recommendation Model for Cyber-Physical Systems." *IEEE Transactions on Emerging Topics in Computing* 1(2): 384–393. doi:10.1109/TETC.2013.2274044.

Geng, Shaofeng, Xiaoxi Guo, Jia Zhang, Yongheng Wang, Renfa Li, and Binghua Song. 2018. A Prediction Method Based on Complex Event Processing for Cyber Physical System. In: *IEEE Communications Magazine*, edited by L. Zhu and S. Zhong. Singapore: Springer Nature. doi:10.1109/MCOM.2018.8419191.

Gunes, Volkan, Steffen Peter, Tony Givargis, and Frank Vahid. 2014. "A Survey on Concepts, Applications, and Challenges in Cyber-Physical Systems." *KSII Transactions on Internet and Information Systems* 8(12): 4242–4268. doi:10.3837/tiis.2014.12.001.

Guo, Yanxiang, Xiping Hu, Bin Hu, Jun Cheng, Mengchu Zhou, and Ricky Y.K. Kwok. 2017. "Mobile Cyber Physical Systems: Current Challenges and Future Networking Applications." *IEEE Access* 6: 12360–12368. doi:10.1109/ACCESS.2017.2782881.

Gupta, Sahil, Shalini Pandey, K.K. Shukla, and K.K. Shukla. 2015. "Comparison Analysis of Link Prediction Algorithms in Social Network." *International Journal of Computer and Applications* 111(16): 27–29. doi:10.5120/19624-1502.

Han, Meng, Zhuojun Duan, and Yingshu Li. 2017. "Privacy Issues for Transportation Cyber Physical Systems." In: *Secure and Trustworthy Transportation Cyber-Physical Systems*, edited by Y. Sun and H. Song, 67–86. SpringerBriefs in Computer Science. doi:10.1007/978-981-10-3892-1_5.

Harel, Yaniv, Irad Ben Gal, and Yuval Elovici. 2017. "Cyber Security and the Role of Intelligent Systems in Addressing Its Challenges." *ACM Transactions on Intelligent Systems and Technology* 8(4): 1–12. doi:10.1145/3057729.

Kashima, Hisashi, and Naoki Abe. 2006. "A Parameterized Probabilistic Model of Network Evolution for Supervised Link Prediction." In: *Sixth International Conference on Data Mining (ICDM'06)*, Hong Kong, 10.

Lee, Jay, Behrad Bagheri, and Hung An Kao. 2015. "A Cyber-Physical Systems Architecture for Industry 4.0-Based Manufacturing Systems." *Manufacturing Letters* 3: 18–23. doi:10.1016/j.mfglet.2014.12.001.

Li, Xu, Rongxing Lu, Xiaohui Liang, Xuemin Shen, Jiming Chen, and Xiaodong Lin. 2011. "Smart Community: An Internet of Things Application." *IEEE Communications Magazine* 49(11): 68–75. doi:10.1109/MCOM.2011.6069711.

Liu, Yang, Yu Peng, Bailing Wang, Sirui Yao, Zihe Liu, and A Concept. 2017. "Review on Cyber-Physical Systems." *IEEE/CAA Journal of Automatica Sinica* 4(1): 27–40.

Liu, Zhong, Dong Sheng Yang, Ding Wen, Wei Ming Zhang, and Wenji Mao. 2011. "Cyber-Physical-Social Systems for Command and Control." In: *IEEE Intelligent Systems*, edited by Daniel Zeng, vol. 26, 92–96. IEEE. doi:10.1109/MIS.2011.69.

Lu, Hui-Chao. 2012. "Porous Materials." *Journal of Porous Materials* 53(619): 743–747. doi:10.9773/sosei.53.743.

Lu, Yiding, Yufan Guo, and Anna Korhonen. 2017. "Link Prediction in Drug-Target Interactions Network Using Similarity Indices." *BMC Bioinformatics* 18(1): 1–9. doi:10.1186/s12859-017-1460-z.

Meenalaxmi, P.S., K. Abinaya, and S. Suseela. 2018. "Social Impact of Cyber Physical System." *International Journal of Latest Trends in Engineering and Technology*: 113–117.

Negenborn, R.R., and J.M. Maestre. 2014. "Distributed Model Predictive Control: An Overview and Roadmap of Future Research Opportunities." *IEEE Control Systems* 34(4): 87–97. doi:10.1109/MCS.2014.2320397.

Ning, Huansheng, Hong Liu, Jianhua Ma, Laurence T. Yang, and Runhe Huang. 2016. "Cybermatics: Cyber-Physical-Social-Thinking Hyperspace Based Science and Technology." *Future Generation Computer Systems* 56: 504–522. doi:10.1016/j.future.2015.07.012.

Rajkumar, Ragunathan, Insup Lee, Lui Sha, and John Stankovic. 2010. "Cyber-Physical Systems: The Next Computing Revolution." In: *Design Automation Conference*. doi:10.1210/en.2014-1673.

Shakshuki, Elhadi M., Haroon Malik, and Tarek Sheltami. 2014. "WSN in Cyber Physical Systems: Enhanced Energy Management Routing Approach Using Software Agents." *Future Generation Computer Systems* 31(1): 93–104. doi:10.1016/j.future.2013.03.001.

Sheth, Amit, Pramod Anantharam, and Cory Henson. 2013. "Physical-Cyber-Social Computing: An Early 21st Century Approach." *IEEE Intelligent Systems* 28(1): 78–82. doi:10.1109/MIS.2013.20.

Sisinni, Emiliano, Abusayeed Saifullah, Song Han, Ulf Jennehag, and Mikael Gidlund. 2018. "Industrial Internet of Things: Challenges, Opportunities, and Directions." *IEEE Transactions on Industrial Informatics* 14(11): 4724–4734. doi:10.1109/TII.2018.2852491.

Socievole, Annalisa, Floriano De Rango, and Salvatore Marano. 2013. "Link Prediction in Human Contact Networks Using Online Social Ties." In: *Proceedings - 2013 IEEE 3rd International Conference on Cloud and Green Computing, CGC 2013 and 2013 IEEE 3rd International Conference on Social Computing and Its Applications, SCA 2013*, 305–312. doi:10.1109/CGC.2013.55.

Wang, Lihui, Martin Törngren, and Mauro Onori. 2015. "Current Status and Advancement of Cyber-Physical Systems in Manufacturing." *Journal of Manufacturing Systems* 37: 517–527. doi:10.1016/j.jmsy.2015.04.008.

Weth, Christian Von Der, Ashraf M. Abdul, and Mohan Kankanhalli. 2017. "Cyber-Physical Social Networks." *ACM Transactions on Internet Technology* 17(2): 1–25. doi:10.1145/2996186.

Xia, Feng, and Jianhua Ma. 2011. "Building Smart Communities with Cyber-Physical Systems." In: *Proceedings of 1st ACM International Symposium on From Digital Footprints to Social and Community Intelligence*, 1–6. doi:10.1145/2030066.2030068.

Xie, Xing. 2010. "Potential Friend Recommendation in Online Social Network." In: *Proceedings - 2010 IEEE/ACM International Conference on Green Computing and Communications, GreenCom 2010, 2010 IEEE/ACM International Conference on Cyber, Physical and Social Computing, CPSCom 2010*, 831–835. doi:10.1109/GreenCom-CPSCom.2010.28.

Xiong, Gang, Fenghua Zhu, Xiwei Liu, Xisong Dong, Wuling Huang, Songhang Chen, and Kai Zhao. 2015. "Cyber-Physical-Social System in Intelligent Transportation." *IEEE/CAA Journal of Automatica Sinica* 2(3): 320–333. doi:10.1109/JAS.7152667.

Xu, Jiachen, Anfeng Liu, Naixue Xiong, Tian Wang, and Zhengbang Zuo. 2017. "Integrated Collaborative Filtering Recommendation in Social Cyber-Physical Systems." *International Journal of Distributed Sensor Networks* 13(12): 1–17. doi:10.1177/1550147717749745.

Yang, Ai Min, Xiao Lei Yang, Jin Cai Chang, Bin Bai, Fan Bei Kong, and Qing Bo Ran. 2018. "Research on a Fusion Scheme of Cellular Network and Wireless Sensor for Cyber Physical Social Systems." *IEEE Access* 6: 18786–18794. doi:10.1109/ACCESS.2018.2816565.

Yu, Xiao, Pan Ang, Lu An Tang, Zhenhui Li, and Jiawei Han. 2011. "Geo-Friends Recommendation in GPS-Based Cyber-Physical Social Network." In: *Proceedings - 2011 International Conference on Advances in Social Networks Analysis and Mining, ASONAM 2011*, 361–368. doi:10.1109/ASONAM.2011.118.

Zeng, Jing, Laurence T. Yang, Man Lin, Huansheng Ning, and Jianhua Ma. 2016. "A Survey: Cyber-Physical-Social Systems and Their System-Level Design Methodology." *Future Generation Computer Systems.* doi:10.1016/j.future.2016.06.034.

Zhang, Yin, Min Chen, Shiwen Mao, Long Hu, and Victor Leung. 2014. "CAP: Community Activity Prediction Based on Big Data Analysis." *IEEE Network* 28(4): 52–57. doi:10.1109/MNET.2014.6863132.

Zheng, Xu, Zhipeng Cai, Yu Jiguo, Chaokun Wang, and Yingshu Li. 2017. "Follow but No Track: Privacy Preserved Profile Publishing in Cyber-Physical Social Systems." *IEEE Internet of Things Journal* 4(6): 1868–1878. doi:10.1109/jiot.2017.2679483.

13

Manufacturing in the Future as a Cyber Physical System – An Overview of Technology to Address Challenges to Scale in Engineer–Secure–Run

Thanga Jawahar, Subhrojyoti Roy Chaudhuri, Swaminathan Natarajan, Harish Mehra and Jitesh Vaishnav

CONTENTS

Organization of the Chapter

Section 1 deals with terms and terminologies for the user to understand the chapter. **Section 2** introduces the concept of cyber physical systems. In **Section 3**, we outline some of the top software engineering challenges towards the engineering of control systems for Industry 4.0-based manufacturing plants, followed by challenges related to the security of cyber physical production systems. **Section 4** describes opportunities for creating the next-generation engineering approaches that mitigate these challenges. **Section 5** presents some next-generation solution approaches, including architecture for cognitive control, and a Domain-Specific Engineering Platform. **Section 6** concludes the chapter.

13.1 Terms and Terminologies

Cyber Physical Systems: Systems which integrate computation with physical processes, the behaviour of which is determined by both the physical and the computational parts of the system.

Context-aware: The capability of adapting behaviour to contextual factors.

Operational Technology (OT): Hardware and software dedicated to detecting or causing changes in physical processes through direct monitoring and/or control of physical devices.

Threat Vector: A path or tool used to attack the target.

Domain-Specific Language: A specification language specialized to a particular application domain, incorporating terms, concepts, operations and patterns common to the targeted class of applications in that domain.

Digital Twin: A software model of the behaviour of a physical entity, which can be executed in parallel with the physical entity to mirror its state, and continuously learns its behaviour. Digital Twins can be used in predictive mode, to evaluate the impact of proposed actions on system behaviour.

13.2 Introduction

Manufacturing industries are witnessing significant technological advances, geared towards enhancing their efficiency and productivity. Cyber Physical

Systems (CPS) are playing a central role in creating a paradigm shift in the workings of the manufacturing plants. Deployment of CPS enables information from all related perspectives to be monitored and used, with close collaboration, coordination and synchronization between the factory plant systems and computational space [1]. This leads the manufacturing industries to transformation into Industry 4.0.

The centrepiece of this emerging paradigm is the control system, which incorporates not only the operational model of the factory plant systems, but also the relevant domain knowledge used to determine the plant system configuration and operational model. However, there are many challenges to realizing control systems.

13.3 Challenges in Engineering Cyber Physical Systems

This section identifies some of the challenges in engineering CPS, both software engineering challenges and security challenges.

13.3.1 Challenges to Software Engineering

Although software engineering practice has been enriched with new methodologies, such as Agile, to improve the engineering process, there is a clear gap that specifically addresses domain-specific engineering challenges, e.g. the engineering of software for Industry 4.0. The challenges for engineering control systems for Industry 4.0 can be categorized as follows:

13.3.1.1 Scale of Engineering

- Much effort is being spent on identifying stakeholder goals and processes and the required set of subsystems and technologies that the control software needs to integrate and orchestrate.
- The synthesis of the system objectives and their processes into a control software design, that ensures achievement of such processes, is a highly effort-intensive task. For example, the design of large complex systems requires investment of hundreds of full-time equivalents (FTE) to design their control software. The tasks also require highly skilled software architects with past experience of building control software for similar systems.
- For the implementation of control software, there are many available frameworks and protocols, such as SCADA packages and PLC technologies. A complex cyber physical system typically uses multiple implementation technologies and protocols (~10 to 20) to

implement their control software. It becomes a nightmare to ensure the integration of all parts of the control software and requires extensive testing and verification through a dedicated commissioning phase.

13.3.1.2 Agility and Adaptability to New Technological Advances

- It is not enough to engineer and deploy the control software; they need also to be constantly reconfigured and adapted to ensure operational efficiency of the system through constant planning, adaptation of new technology, different situations and so on.
- With Industry 4.0, the paradigm of fixed, highly predictable behaviour by design, no longer holds. Internet-of-Things (IoT) technologies enable easy introduction and removal of devices from the plant system network. Robotics introduce smart machines that exhibit complex and flexible behaviours, and incorporate planning and learning functions, so that they are no longer limited to a fixed pattern of operation. Analytics make it possible to detect complex situations, whereas AI, Cloud services and increasing computational capacities make it possible to implement intelligent and sophisticated responses to these situations.

13.3.1.3 Self-Organization Through Learning from History and Collaboration

- One of the key objectives of the next-generation manufacturing industries is to improve human–machine collaboration and to enable machines to learn from explicit and implicit data. This requires the control system to be able to learn from all sources of knowledge, such as previous job descriptions or implicit gestures from human collaborators.
- Technologies, such as machine learning and deep learning, already allow control systems to move away from making fixed decisions to exhibiting context-aware strategy executions. However, a key challenge with the learning technologies is to enable their learning, by providing them with the training data, so that they guarantee the required reliability and safety, demanded by manufacturing plants which implement them.

13.3.2 Security of Cyber Physical Production Systems (CPPS)

It is important to understand the CPS ecosystem, its associated assets and its criticality, to address the security associated with it. These assets can be secured of vulnerabilities. The CPS will have a very complex ecosystem, involving huge number of assets, cutting across sensors to the cloud.

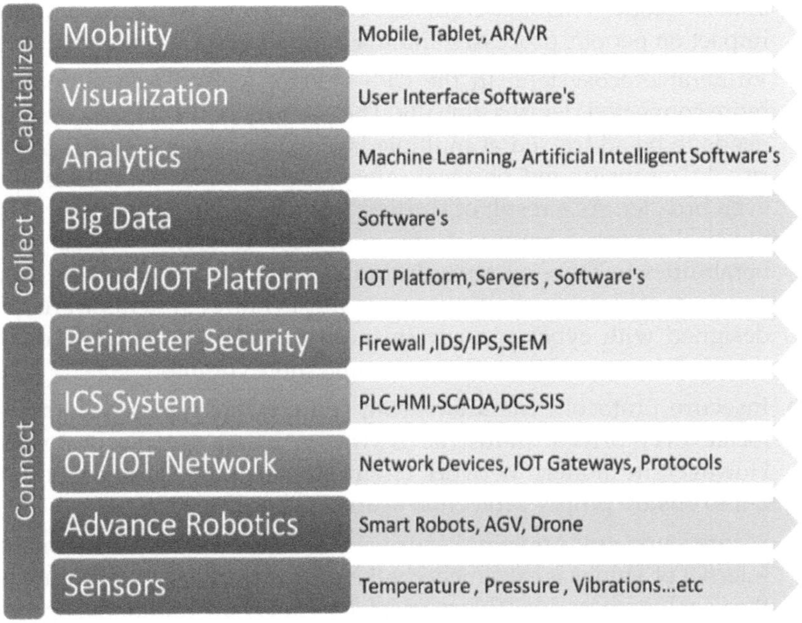

FIGURE 13.1
CPS ecosystem assets

To represent the assets' view in the CPS ecosystem, we try to divide the CPS ecosystem into three areas, like connect, collect and capitalize, as represented in Figure 13.1. The assets represented in the connect area are complex in category and will have greater impact on the CPS landscape. The typical assets under connect are sensor, advance robotics, an Operational Technology (OT) network, edge devices, OT system and security devices, which will be associated more with the physical layer of manufacturing or the supply chain network.

The collect area is moderately complex in nature and has assets like IoT platforms, servers, software, etc. These assets will be associated more with cyber systems.

The capitalize area is of low complexity and has assets like software for analytics, user interfaces and mobile devices.

It is very clear that Cyber Physical System (CPS) makes the operational technology very handy to operate. It also becomes very easy for machines to take decisions on various conditions, using the in-built AI. But, with these advances in the current operational environment, CPS brings some cyber security-related challenges, such as those listed below:

- Larger number of devices and their connectivity: In the CPS ecosystem, it is expected to have a greater connectivity to the supply chain network, where physical systems will interact with cyber systems at a faster mode. This is the greatest challenge and will have a direct impact on people, processes and machine safety.

- Vulnerable ecosystem: In the CPS system, we will see more and more connected devices globally. The security of the CPS ecosystem needs to be addressed at multiple layers, like IT security, OT security, IoT security and physical safety, which makes this landscape even broader. As a result of shifting from closed to connected cyber physical systems, manufacturers will need to handle typical vulnerabilities in these systems. In industrial environments, this may pose a considerable challenge, since most control systems were not designed with cyber security in mind and thus vulnerabilities in their hardware are becoming more and more common.
- Insecure protocols: Manufacturing components typically communicate over private industrial networks, using specific protocols. However, in modern network environments, these protocols often fail to ensure proper protection against cyber-threats
- Security updates: Applying security updates to control systems on the Internet of Things is extremely challenging, since the typical nature of the user interfaces available to users does not allow traditional update mechanisms. Securing those mechanisms is, in itself, a daunting task, especially considering Over-The-Air updates. In OT environments, in particular, applying updates may be challenging since this operation needs to be scheduled and performed during downtime.
- Unused functionalities: Industrial machines are designed to offer a large number of functions and services, many of which may not be necessary for operation. In industrial environments, machines or their selected components often have access to these unused functionalities, that may considerably expand the potential attack area and become gateways for the attackers.
- Supply chain complexity: Companies that manufacture products or solutions are very rarely able to produce every part of the product itself and usually need to rely on third parties' contributions. Developing technologically sophisticated products results in an extremely complex supply chain with a large number of people and organizations involved, thereby making it highly demanding in terms of management. Not being able to secure its source means not being able to ensure product security, which is only as secure as its weakest link.
- People awareness: Adoption of new technologies means that factory workers and engineers have to work with new types of data, networks and systems in novel ways.

They are unaware of the risks associated with gathering, handling and analyzing that data and can thus become an easy target for attackers.

13.3.2.1 Threats and Attacks on CPPS

Attacks on industrial control systems (ICS) have increased to a large extent, and ICS have become soft and easy targets for cyber hackers during the past decade. Today, the attacker gets more space and multiple ways to break into CPS because of the increased connectivity.

Threats and attacks at various layers and components of the ecosystem are shown in Figure 13.2. Some of them are explained below.

- Denial-of-Service (DOS) Attack:

  If the requester is requesting any service but the request is not able to reach to the service provider within the given timeframe, this is referred to as a denial-of-service. This is one of the easiest possible attacks to perform on the cyber physical system. Denial-of-service simply means being denied from giving services. So, if any attacker gains success in choking the CPS network and prevents the service requests from reaching up to their destination, successful denial of service has been achieved.
- Man-In-the-Middle (MIM) Attack

  There are many ways through which the man-in-the-middle attack can be performed on cyber physical systems. Man-in-the-middle is an attack where an attacker sitting in the network can read/write/ modify the messages, as in CPS, when control points are set from the

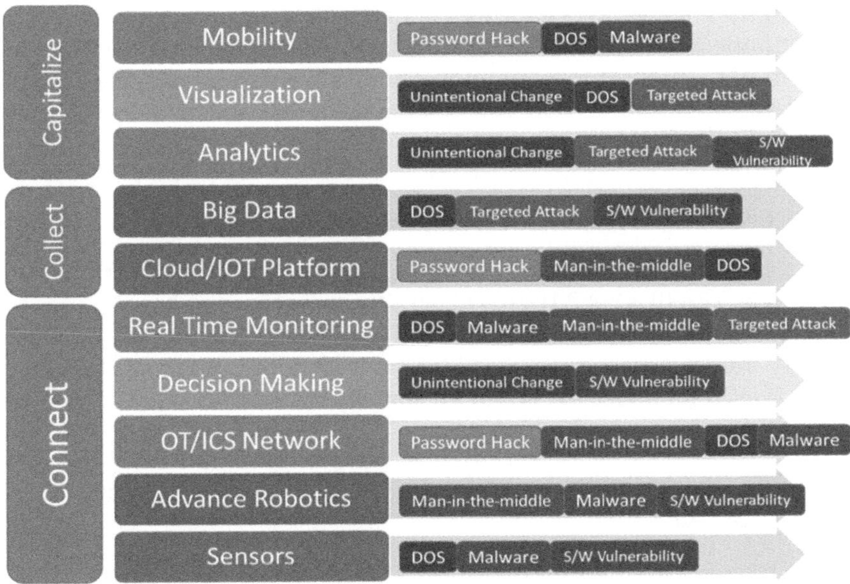

FIGURE 13.2
Threats and attacks relating to each CPS ecosystem layer

cloud after doing various analysis on live data. So, if any attacker, at the time of giving the control command, modifies that command, it can create a devastating impact on the control system.

- Social Engineering Attacks

  Working with the cyber physical systems is quite different from working in the legacy control system environment. All stockholders must be aware of the importance of data they are handling in day-to-day operations, the importance of their own credentials and the importance of the critical information which they have as a part of the cyber physical system. In the absence of proper training, operators, engineers and all relevant stakeholders may end up becoming a victim of a social engineering attack, thereby disclosing the information.

- Device Hacking

  As we know in CPS there comes a use case where the plant maintenance manager can control and operate the machine from his mobile phone from his house. The threat vector involved in this use case is whether the mobile device gets hacked. In this case, the hacker will be able to operate the machine from the mobile and can cause damage.

13.3.2.2 Opportunities in Securing Cyber Physical Systems

Much of the engineering challenges outlined above also provide significant opportunities towards their mitigation, through research and development on tools and technologies, methodologies and new knowledge. We will discuss these opportunities with examples of solutions, already demonstrating next-generation approaches.

13.3.2.3 Security Measures, Methodologies and Standards

For a robust cyber security system, there are a few industry standards such as NIST, ISO 27001, OSAP, IEC62443, which can be leveraged.

NIST is typically used for IT and OT security. ISO27001 is a corporate security framework whereas IEC 62443 is the standard used for OT security. OSAP is used for applications and IoT for services security.

13.4 Opportunities for Intelligent Operations and Automation

The essence of the Industry 4.0 paradigm shift is the opportunities for intelligence and automation in operation, offered by a suite of technologies:

- IoT offers opportunities for extensive instrumentation of operations, mostly by using relatively low-cost networked sensors. By gathering the right information, combining information from multiple sensors and adding soft sensors (derivation or estimation of information from available inputs), it is possible to build up a detailed picture of the state and behaviour of the plant system. IoT also makes it easy to dynamically integrate new devices into the plant system network and remove them, especially if their interaction with other devices primarily involves exchange of information rather than materials. This creates the possibility of rapid reconfiguration of plant systems to provide variant functionality, subject to any limitations imposed by physical connectivity.
- Analytics processes the gathered information, to obtain information about key system operating parameters, to detect situations of interest (e.g., using trend information to detect faults and threats before they lead to failures) and also to derive insights into possible influencers of observed behaviour patterns. Thus, it provides knowledge about current behaviour and possible future states of the system, and also helps to refine the understanding of the system that we have from domain knowledge. The ultimate analytics enabler for plant systems is a full Digital Twin of a device, which incorporates a behavioural model of the device that asymptotically converges to the actual behaviour, evolves as the device's behaviour changes, and can be used for predictive "what-if" decision-making in advance of operations.
- Machine learning improves decision making by observing correlations between situations, decisions and outcomes, creating the possibility of continuous improvement of control strategies by observing their effectiveness in the actual situation. This offers three advantages over fixing control strategies at design time:
    - Fixed strategies are derived, based on domain knowledge, and are therefore only able to account for situational factors that have been explicitly identified and codified. Often in real-world situations, there are many more factors that may be operating, and, moreover, these interact in complex ways. Learning can take a wider range of situational variables into account in decision making and can accommodate deep and complex combinations of these. This is why it is particularly effective on problems, such as grasping objects, where the factors involved, and their interplay are hard to codify into decision rules.
    - Fixed strategies are based on design-time information about the system-of-interest. Learning operates at run time and is therefore able to take into account any variations introduced by the implementation and in the specific deployment environment.

    - Depending on the specific techniques employed, learning may be able to adapt to situational changes without requiring design changes.

  One major limitation of learning is that it may not be able to deal with situations that have never been seen before (including during training). This can be a problem in embedded systems, where some critical failure situations are known as part of domain knowledge but may not be easy to introduce into training data or encountered in normal operations. For this reason, in plant systems control, it is desirable to complement the learning with the use of domain knowledge to ensure acceptable behaviour in critical situations. We will see later on how this can be accomplished in cognitive control.

- Robotics and smart machines, such as drones, allow flexible automation of particular tasks, including some tasks that may be tedious, difficult or dangerous for humans. Some robots can perform a wide range of tasks, and many robots can make behavioural decisions, based on task planning. The plant system must be able to accommodate this unpredictability within its operational model. A major challenge (not addressed in this chapter) arises when robots and people interact in plant system operations: we need to ensure safe and effective collaboration between them.
- Cloud computing and increasing computational power enables the incorporation of sophisticated algorithms for situation appraisal and decision making, as well as external services and information sources, to be incorporated into the plant control system.

13.5 Next-Generation Solution Approaches

In this section, we present some of the next-generation approaches and ideas to enable the development of engineering cognitive control systems for Industry 4.0-based manufacturing plants.

13.5.1 A Conceptual Architecture for Next-Generation Solutions

The proposed paradigm of cognitive control is knowledge-centric: it incorporates domain and contextual knowledge as part of the control system, it builds and refines this knowledge, based on analytics of information gathered from plant system functioning, and it exploits this knowledge to direct, constrain and update the functioning of the operational control system. Figure 13.3 shows reference architecture for the cognitive control paradigm.

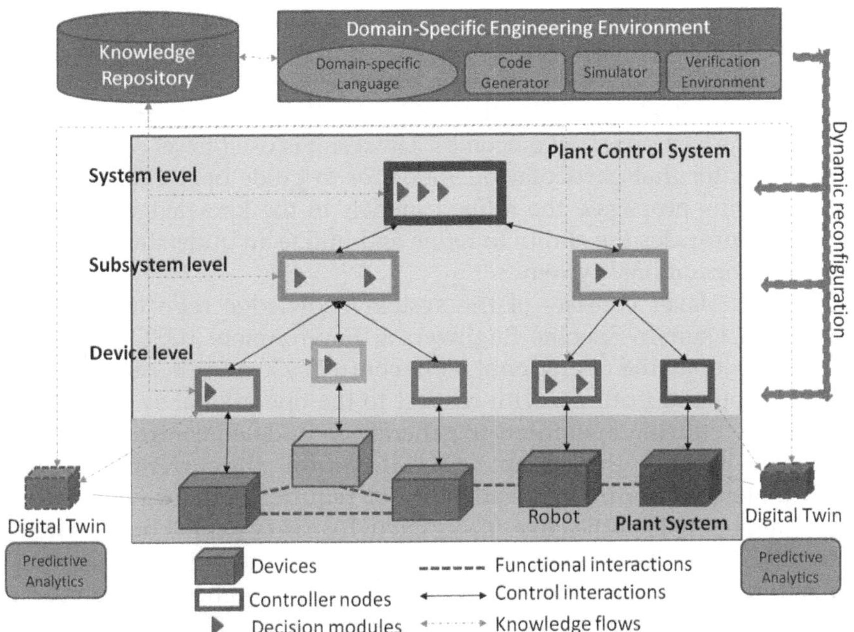

FIGURE 13.3
Architecture for cognitive control

The architecture consists of three layers that operate on different time scales and bring knowledge to bear in handling successively more challenging situations. The innermost layer is the operational plant control system, shown as a grey box in Figure 13.3. This consists of a hierarchy of controllers, similar to the traditional control paradigm. Controllers may contain one or more decision modules, which may include learning capabilities. They provide increasingly more intelligent decision making by observing the outcomes of previous decisions made in similar contexts. This learning may be propagated back to the knowledge repository, and, as we shall discuss later, the decision making is constrained by knowledge-based behaviour envelopes. Decision modules may include functions such as planning and scheduling, that are typically part of the Distributed Control System (DCS) layer, but also action choices, while performing operational tasks and handling fault scenarios at the device and subsystem (SCADA) controller levels. This layer provides the capability to learn from previous experience.

The middle layer consists of Digital Twins, that operate near-inline, with respect to plant operations (i.e., they observe operations and occasionally provide directive inputs). They learn the operational model of particular devices and subsystems and have associated predictive analytics capabilities that identify the best ways to handle particular situations. Digital Twins are seeded with initial models, based on domain knowledge, and the model is continually refined, based on the monitored data gathered during operations. This refined model can be used for "what-if" simulations, prior to executing particular operations (such as a production run), to explore a series of options for dealing with that operational situation and identifying the one with the best projected outcomes. The result is used to set configuration parameters for that particular operation, or to guide operational strategy. Digital Twins propagate the refined models to the knowledge repository. This layer provides the ability to refine and utilize an understanding of the particular operational system.

The outer layer consists of the system knowledge repository and an associated Domain-Specific Engineering Environment (DSEE), which is used to generate the operational plant control system. Whereas this layer normally operates offline, with respect to the operational system, it may be triggered during operations to generate an updated control strategy to deal with situations that cannot be handled within the current operational model, such as resource unavailability or requirements for new features which are not part of the current system (which may exploit devices newly introduced into the system and not covered by the current operational model). In such cases, the DSEE automatically searches the knowledge base to determine whether it is possible to create an updated control strategy that is capable of meeting the requirements or to provide some acceptable handling of the scenario. If successful, the updated control strategy produced is automatically simulated and verified. The output from the DSEE involves a dynamic reconfiguration of the plant system and updates to its controllers, including possibly additional controllers to manage the newly introduced device and coordinate its interactions with other devices. It should be noted that the reconfiguration may include manual actions to establish the required physical connectivity. This layer provides the ability to generate and update control strategies based on available and gathered knowledge.

In cases where the DSEE automation capabilities are unable to find a solution to the situation, it is up to the engineer to determine how to handle the scenario, perhaps by augmenting the knowledge and using the DSEE to develop an updated control strategy.

The knowledge repository contains structured representations of the particular system configuration, its operating environment, desired operational outcomes and applicable domain knowledge. The structure and organization of the knowledge repository will be discussed in a later section.

It can be seen from the above (and the discussions in the rest of this chapter) that the cognitive control paradigm involves the following key themes:

- Information gathered with IoT is converted to usable experiential knowledge with analytics, and contextualized using Digital Twins, enabling a deeper understanding of system function.

- Domain knowledge and problem-specific knowledge (requirements, desired outcomes, design schemas, system operating model), which are primarily tacit in the minds of engineers in the current paradigm, are captured as explicit structured knowledge in the knowledge repository, and then refined and contextualized, based on observed behaviour. In some simple cases, this knowledge can be used to automatically modify the engineering design, whereas, in more complex situations, we rely on engineers and provide them with an engineering environment that enables them to leverage the structured knowledge in the repository.
- The operational plant control system is kept relatively simple, particularly because edge devices may be relatively limited in their computational capacity. The more sophisticated tooling, such as Digital Twins and the DSEE, is not inline, with respect to the basic plant system operations. Even sophisticated decision-making and learning capabilities are optional in controllers, depending on what is appropriate to the particular device and application. Where adequate computational capacity (and time for decision making) is available locally or on the cloud, controllers can incorporate sophisticated decision algorithms and situation-detection algorithms (e.g., pattern recognition algorithms – this is not shown in the diagram).

13.5.2 Engineering Platforms

Domain-specific engineering platforms are developed specifically to mitigate the engineering challenges by providing:

- A standard platform to design and develop control systems, following standard control software architecture patterns, which needs to be contextualized only for different complex cyber physical systems. Hence, it eliminates the need to reinvent architecture and the design of control software from scratch for every system.
- Use of domain-specific languages (DSL) and Controlled Natural Languages (CNL), raising the levels of abstraction for specifying the design of control systems. This ensures that the design of the various parts of the control software is created using a standard language, ensuring uniformity in the usage of concepts and vocabularies from the domain of interest. It eliminates the possibility of any inconsistencies in the control software design, created across distributed teams.
- Explicating domain knowledge to domain-knowledge repositories. Such domain knowledge can be reused and composed by domain specialists and systems architects to come up with process definitions that a complex cyber physical system should implement.

It makes the synthesis of such processes to the design of corresponding control software transparent and automated [2].

- Usage of Model-Driven Engineering, to eliminate manual coding by auto synthesizing implementations and their testing and simulators.
- Continuous monitoring of the behaviour of the control software against its intended design during its operation, to detect anomalies across multiple maintenance and evolution cycles.

13.5.3 Domain-Specific Engineering Platform Architecture

Figure 13.4.

13.5.4 Domain Knowledge Repositories

Knowledge is at the heart of cognitive control. The knowledge repository includes relevant domain knowledge, knowledge about the particular plant system, its components and operating environment, knowledge about desired behavioural outcomes, and contextual knowledge about behaviour of the deployed devices and the plant system in their operating environment. The challenge is to create structured representations of all this information as a knowledge repository, that can be used by the DSEE platform and other tools.

Knowledge about entities, such as devices and how they organize into subsystems and an overall system, can be expressed using semantic web technologies such as Resource Description Framework (RDF), as shown in Figure 13.5. A key idea is that we need to capture information, not only about entities, their properties and their relationships to other entities, but also about the *capabilities* that they provide. A capability description identifies

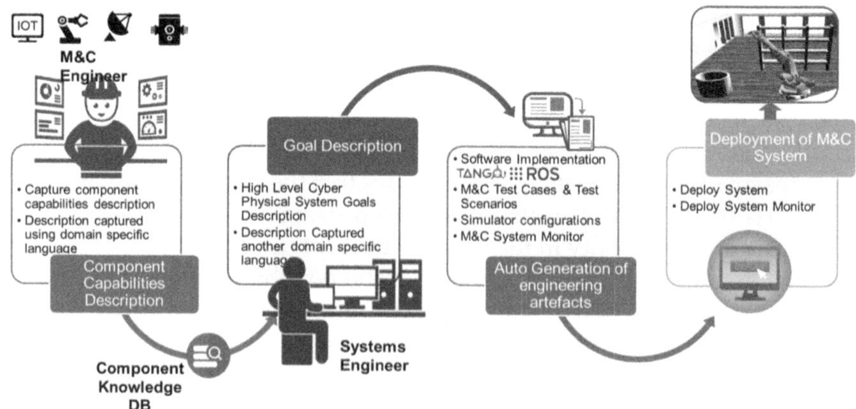

FIGURE 13.4
Domain-specific engineering platform usage workflow

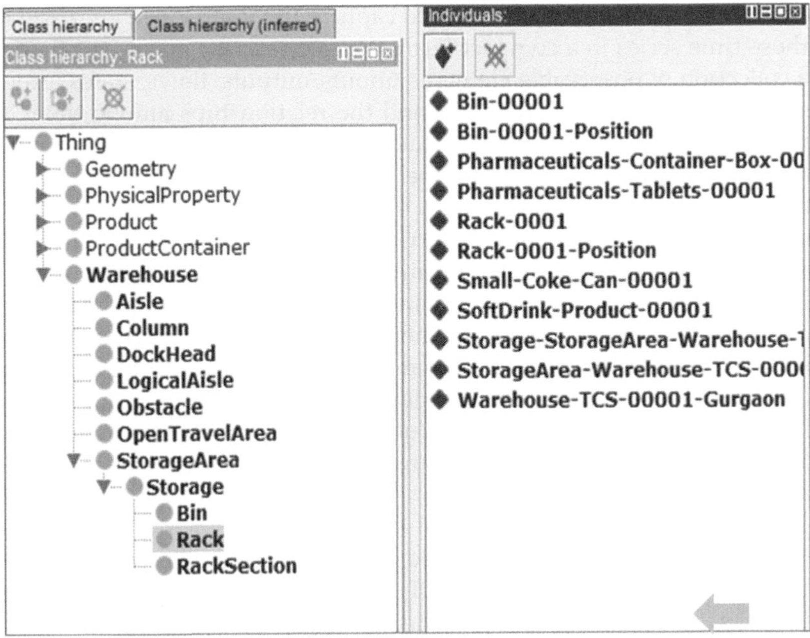

FIGURE 13.5
Use of Semantic Web Technologies to capture domain knowledge

the function performed by an entity, associated features (add-on and variant functionality, including security and safety features), as well as the service interfaces for using the capability, and associated quality characteristics (performance, availability, etc.) and resource dependencies. Capability descriptions permit selection, among alternative resources and devices, to find one that meets the particular contextual needs. The set of entities and associated characteristics is specific to each domain, so we must build a domain knowledge repository of concepts and entities of interest in that domain. The model includes not only concepts relating to the plant system, but also those pertaining to the environment of the plant system, e.g., a warehouse operations system would include a description of warehouse concepts, including its physical layout.

In addition to capturing the characteristics of the entities, we need to also capture the process patterns involved in achieving goals and outcomes. This involves the description of procedures and algorithms, so it is best expressed in a language. Since the procedures operate on domain objects, ideally the language should be specific to that particular domain. The domain-specific languages (DSLs) described in the next section are the ideal vehicle with which to capture the process patterns in the domain.

Finally, we also need structured representations of behaviour, including the insights resulting from analytics. Whereas monitoring produces streams of time-series data, we need a way to capture the knowledge encapsulated in those time series in a compact form. We can describe behaviour in terms of a collection of observable elements (inputs/outputs, flows, states, properties, structure, events, interactions) and the relationships among them. For example, transactions and physical laws define relationships among inputs, states and outputs. The response time of a module can be captured in terms of a range of values associated with the "response time" property. We can create similar specifications of the arrival rates of inputs, the variety of possible values for a given input, constraints on legal combinations of values for multiple outputs etc. Insights from analytics can now also be expressed as relationships among observable elements. It should be noted that the vocabulary of observable elements is domain specific and is part of the collection of concepts captured in the RDF model.

The above schema, for creating structured representations of entities and concepts, processes and behaviour, provides the mechanisms needed for the knowledge repository:

- Domain knowledge is captured in terms of a domain vocabulary of entities and concepts in RDF, process patterns expressed using DSLs, and behavioural knowledge about expected relationships among observable elements.
- Plant system and environment knowledge is expressed as a system model, defined over the RDF entities and vocabulary, and associated properties and behavioural relationships (for the behaviour of the environment).
- Desired behavioural outcomes are expressed as behavioural relationships over the properties in the RDF model.
- Contextual knowledge about the observed plant system behaviour is expressed as behavioural relationships over the properties of the devices and subsystems in the plant system model.

Thus, we can create a knowledge repository that expresses all the information and knowledge relating to the plant system in a uniform domain-specific structured representation.

13.5.5 Domain-Specific Languages: Raising Engineering Specifications to the Level of Knowledge

Currently, control software and control system solutions are implemented by engineers in a patchwork of technologies, including languages, such as Java and C++, PLC tools, DCS packages, SCADA packages and auxiliary tools and technologies including loggers, archives, alarm handling tools, databases, presentation tools, etc. Whereas each individual tool improves productivity and provides powerful capabilities, there is no integrated view of the overall solution (except in the engineers' minds). The operational model of the control system cannot readily be extracted from this patchwork.

In contrast, from the domain point of view, the operational model of any controller is pretty straightforward. Whereas particular continuous control algorithms, situation detection modules and decision modules may encapsulate complex logic within them, the overall discrete control logic is relatively straightforward, implementing well-known patterns in the domain for orchestrating behaviour, achieving particular functional and quality outcomes, and handling various situations.

The take-home message is that it is highly desirable to express the operational model for the control system in domain-specific terms, where the patterns involved would be immediately clear. Figure 13.6 shows an example of a DSL for the control systems domain (which is useful to express the operational logic of device and subsystem controllers), whereas Figure 13.7 shows an example of a DSL for expressing business workflows in the warehouse operations domain; such DSLs are needed in the DCS layer.

13.5.6 Model-Driven Engineering: Automated Generation of Life Cycle Artefacts

Once the desired behaviour is expressed using these DSLs, implementations in various technologies can be automatically generated from these descriptions. In effect, we are raising the level of solution engineering from the

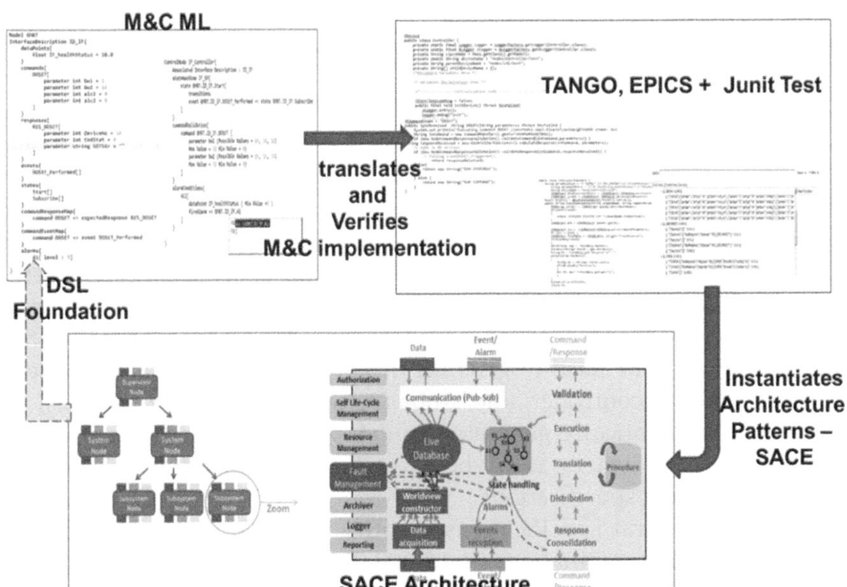

FIGURE 13.6
Control systems DSL (M&C ML)

```
Resource: Mobile_Robot
Description: "Mobile_Robot" consists of multiple components. It uses a "Simple_Mobile_Chassis_With_Two_Wheels" and a "Simple_Arm_With_Gripper".

Abstract Solution:

The "Mobile_Robot" should "Move To Location" of "Rack".
The "Mobile_Robot" should "Move Arm" close to the "SmallCan".
The "Mobile_Robot" should "Open Gripper Palm" next to the "SmallCan".
//I think in the previous statement, the mention of the SmallCan can be omitted.
//If it is not mentioned, then the reference to the same should be picked up from the previous statement.
The "Mobile_Robot" should "Close Gripper Palm" to hold the "SmallCan".
The "Mobile_Robot" should "Move To Location" of "Bin".
The "Mobile_Robot" should "Open Gripper Palm" at location "Bin".

The "Mobile_Robot" should "Close Gripper Palm" after putting the "SmallCan" on the "Bin".

//In the statement above, the last part starting from "after putting..." is redundant.
//This is because the placing of the can at the bin was already covered in the previous steps.

Solution 1 : "Rack" = "Rack_0001" , "Bin" = "Bin_00001" , "SmallCan" = "Small_Coke_Can_00001"
```

FIGURE 13.7
Domain-specific language for expressing task workflow

implementation technology level to the domain knowledge level. This has huge impacts on life cycle costs. Moreover, because the operational model is expressed in the same terms as the domain knowledge, it becomes possible to automatically synthesize parts of the operational model by combining the requisite patterns from the domain knowledge and selecting the appropriate resources to provide the required capabilities for performing the tasks involved in the patterns.

If the right information is present in the domain knowledge repository and in the DSL- specification of the operational model of controllers, it is possible to automatically generate simulations of devices and subsystems and to automate verification.

For hardware devices, if the associated behavioural specifications capture their responses to particular inputs (both in terms of state changes and outputs), then that can be used to build stub simulations of hardware devices. For controllers, DSL specifications include both their control (parent) interfaces and their interfaces with child devices or controllers (and peer interfaces, if any). If there are behavioural specifications of partner system behaviour associated with those interfaces, or if there are detailed operational model specifications available for the partner systems, those specifications can be used to generate driver implementations corresponding to the parent controller, and stubs for the other interfaces. Thus, it is possible to automatically generate simulators, both for individual devices and for the plant system as a whole, from the knowledge repository and operational model specifications.

If the behavioural specifications identify the space of values for each input (and valid combinations), and, ideally, if those inputs are grouped into equivalence classes, it is possible to automatically generate verification cases that run the devices or plant systems through large numbers of scenarios, and to verify whether the plant system meets the desired outcome specifications.

In short, as we increasingly capture knowledge about the plant system, its environment, and relevant domain knowledge in structured form, it will become possible to achieve increasingly higher levels of automation of the entire engineering life cycle, even in the presence of intelligent system elements and features, such as learning. With the combination of behaviour envelopes and automated simulation and verification, plus Digital Twins to learn the actual contextual behaviour of the plant system, it might even be possible to achieve higher levels of reliability than with the current paradigm.

13.5.7 Continuous Monitoring and Digital Twins

Control systems engineering includes not only the specification of actions to be taken by the control system, but also how the physical system is expected to respond to these actions. In other words, the engineering of plant systems includes expectations of how the controlled plant system will behave in different situations – it is this identification of target behaviour that drives the entire engineering process.

By including these specifications of expectations as part of the engineering, we can generate tools that monitor the actual behaviour of the system, compare it with these expectations, and flag deviations for refinement of the engineering model. Continuous monitoring is an offline approach for this, that analyzes log files and monitoring data against expected behaviour, and flags deviations for analysis by engineers.

Digital Twins are part of the actual operational environment, and operate in parallel with the plant system, observing the streams of monitoring data and building/refining models of system behaviour. Digital Twins can be used for predictive analysis.

13.6 Conclusion

Realizing the factories of the future as complex Cyber Physical Systems is still a very fertile field for research and innovation. The key to realizing these systems will call for becoming part of the journey. A bleeding-edge technological framework is emerging with a comprehensive insight on the needs for the future, where it can be scaled to make huge advances in the field of manufacturing, which is crucial for feeding to the increasing need of the society of the future.

References

1. Lee, Jay, Behrad Bagheri, and Hung-An Kao. "A Cyber-Physical Systems Architecture for Industry 4.0-Based Manufacturing Systems." *Manufacturing Letters* 3, 2015: 18–23.
2. Chaudhuri, Subhrojyoti Roy, et al. "A Knowledge Centric Approach to Conceptualizing Robotic Solutions." In: *Proceedings of the 12th Innovations on Software Engineering Conference (Formerly Known as India Software Engineering Conference)*. AC Med, 2019.
3. Good Practices for Security of Internet of Things in the Context of Smart Manufacturing by Ensia.

Section 4

Advances, Challenges and Opportunities in Cyber Physical Systems Security

14

Cyber Physical Systems Threat Landscape

Suhas Shivanna

CONTENTS

Organization of the Chapter

Section 1 deals with terms and terminologies for the user to understand the chapter. **Section 2** introduces the concept of Cyber Physical Systems. **Section 3.1** gives examples of key security trends and the CPS threat landscape. Opportunities and key solutions for securing Cyber Physical Systems are presented in **Section 3.2**. **Section 3.3** elucidates emerging technologies and opportunities for securing Cyber Physical Systems, and **Section 4** concludes the chapter.

14.1 Terms and Terminologies

- **Information Technology (IT)**:

  The use of systems/devices, like servers, storage and networking, for storing, processing and sending information/data.
- **Operation Technology (OT)**:

  Hardware/software used for monitoring and control of industrial equipment and processes
- **Trusted Computing Group (TCG)**:

  TCG is an industry body that develops open standards and specifications to allow trusted computing
- **Threat Modelling**:

  Threat modeling is a proactive process to identify, evaluate and mitigate potential threats and vulnerabilities

14.2 Introduction

With the increasing adoption of Cyber Physical Systems (CPS) across various industries, like manufacturing, avionics, automobiles, critical infrastructure, defence and oil and gas, securing and safeguarding the Cyber Physical Systems will be very important to protect the interests of the nation and enterprises embarked on the digital transformation journey. In this chapter, we will introduce the key CPS security trends and potential threats to the modern world due to the activities of motivated cyber criminals, and the presence of cyberterrorism and nation state sponsored cyber-attacks. We will cover some of the key security features for Cyber Physical System and the Industry 4.0 smart factory solutions to deal with the modern threat landscape. We will also look at the recommended secure development practices that should be followed when designing Cyber Physical Systems. At the end of this chapter, we will introduce several research opportunities for improving Cyber Physical System security, with respect to some emerging technologies and the changing threat landscape. This chapter will cover the following topics across various sections:

Section 1: Key security trends and CPS threat landscape.

a) Supply chain security challenges.
b) CPS technology and solution architecture and associated security challenges.
c) CPS environment and ecosystem challenges.

Section 2: Opportunities and key solutions for securing CPS.

d) Securing the CPS components and solutions.
   - Hardware Root of Trust in CPS.
   - Secure CPS cryptographic module design.
   - Trusted computing and attestation solution for CPS.
   - Key product security features
e) Secure development methodology for hardware and software powering the CPS.
f) Security by Default and advanced physical security features.
g) CPS secure operations.
   - End-point protection.
   - Segmentation and network isolation.
   - Hardening CPS.
   - Proactive vulnerability management.
   - AI-based security monitoring system.
   - Periodic assessment and audit of CPS.

Section 3: Emerging technologies and opportunities for securing Cyber Physical Systems.

- Quantum computing safe CPS design.
- Blockchain for supply chain and identity management.
- Advanced security hardware accelerators.
- Advanced security analytics for Cyber Physical Systems.
- Industry-specific reference threat models and threat intelligence.
- Converged edge computing.

Section 4: Conclusion

Bibliography

14.3.1 Key Security Trends and CPS Threat Landscape

With rapid digitization and transformation of enterprises and factories, security is increasingly becoming a very important factor in all CPS implementations. Over the past decade or so, cybersecurity has moved on from the world of script kiddies, who hacked into systems for fun and thrills, to motivated cyber criminals, aiming to make fast money by stealing data and identity. With CPS enabling smart grids, critical infrastructure and defence solutions, the current cyber security threat has reached a fever pitch, where wars are now fought in cyber space rather than in the physical world. In the complex world of cybersecurity, the attacks and threats can come from anyone, from a naïve user, clicking on a malicious link or connecting an infected USB, to disgruntled employees, nation state-sponsored groups, cyber criminals, hacktivists and cyber terrorists.

Many of the small-scale attacks go unreported, whereas a few incidents of attacks on CPS have shaken up the industry and taken the world by surprise. A classic example, that illustrates this change in the threat landscape, can be seen in the Iran nuclear plant attack, where multiple centrifugal machines, used to enrich uranium, were crippled using an advanced destructive malware called Stuxnet. The recent German steel plan attack is an example where complex machineries were destroyed using sophisticated malwares designed by highly motivated cybercriminals. With critical infrastructure, like dams and smartgrids, connected to the Internet and controlled by machines, there is increasing threat from nationstate actors, who are rumoured to be creating sophisticated cyber weapons that can cripple any critical infrastructure, using security vulnerabilities and design flaws in critical CPS solutions. An example of this type of attack can be seen with the crippling of the Ukraine electric grid infrastructure, that is rumoured to have been carried out by nation state-sponsored cyber criminals from a neighbouring country. There is also increasing evidence from cyber security intelligence bureaux of a new type of attack called Permanent Denial of Service (PDOS), that is aiming to cripple IT and IoT systems with the intent of infecting and damaging the device into an unrecoverable or 'bricked' state. This kind of attack, also known as 'phlashing', can result in the device becoming unrepairable or only recoverable by returning it to the manufacturing facility. One of the publicly available reports of a PDOS attack used a malware called BrickerBot and is rumoured to have been used to destroy more than one million connected IoT devices, using well-known vulnerabilities and default passwords.

Modern CPS systems, like autonomous cars, have hundreds of embedded systems, connected by various communication technologies, ranging from Bluetooth protocols to cellular networks. It is estimated that the number of lines of code inside such a complex autonomous system is in the tens of millions, and any vulnerability inside the embedded systems or the applications controlling the CPS can cause major catastrophic events, resulting in loss of human life and property damage. In the past few years, many potential attacks, exploiting vulnerabilities, have been demonstrated successfully, and it is safe to assume that, in the modern world, attacks on CPS are not theoretical anymore. With CPS building upon the IoT technology and using varying numbers of embedded systems for measuring and controlling industrial processes, such large-scale attacks cannot only cripple the economy, but can also cause damage to property and even take human lives. It is highly important to look at holistic security practices, that start from securing the supply chain and building security in all the components of a CPS, to protect the CPS from this complex threat landscape. It is also important to look at secure operational practices, that span the complete CPS ecosystem, to deal with motivated cyber criminals and protect the nation, industry, factory and the machinery of the modern digital world. Let us look at some of the above areas in more detail, starting from some security challenges and associated threats, which are critically important for Cyber Physical Systems.

14.3.1.1 Supply Chain Security Challenges

With nation state-sponsored activities, geopolitical tensions and the constant battle for supremacy in the modern digital economy, there is an increasing threat of advanced supply chain attacks, aimed at installing malwares before the system is shipped to a consumer. A supply chain attack can be carried out on the manufacturing floor or in the data warehouses or during transit, using physical access to these devices, and advanced tools and techniques. If these malwares are installed on IoT devices at the hardware and firmware layers of gateways controlling the Cyber Physical Systems, it will become very difficult to detect these advanced malwares and root kits because of the lack of proper malware scanning tools on the lower layers, such as firmware or hardware. With critical infrastructure-based CPS solutions, powered by embedded systems that have complex software, validation of the supply chain integrity, to ensure the state of the device is as expected and not compromised, becomes important in any secure deployment of CPS solutions.

14.3.1.2 CPS Technology and Solution Architecture and Associated Security Challenges

With the ever-changing threat landscape and the use of various technologies and products in the Cyber Physical System solutions, let us look at the main challenges in securing a Cyber Physical System.

A typical CPS has various technologies and sub-systems, starting from the physical machinery, complex control systems, computing devices at the edge, acting as gateways, and communication networks interconnecting the CPS and cloud-based platforms and services. Figure 14.1 illustrates a typical 4-stage CPS/IoT reference architecture, as described by Hewlett Packard Enterprise, one of the leading CPS, IoT and edge computing solution providers in the market. The solution can be split into operation technology (OT),

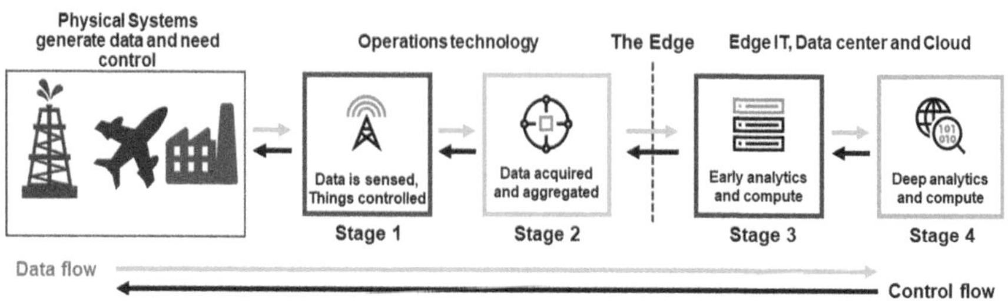

FIGURE 14.1
Typical CPS reference solution architecture.

that comprises hardware and software control systems closer to the physical system and processes usually found in automobiles, on manufacturing floors and in smart grid generation and distributing facilities. Information Technology (IT) consists of the traditional data centre and cloud computing technologies, consisting of server, storage, networking infrastructure and applications. The edge is defined as the boundary between the OT and the IT and is normally the place that is outside the data centre/cloud and closer to the place where you have things or physical devices/processes. The computing infrastructure used near the edge is called the edge computing and supports a combination of wired and wireless protocols to communicate with the OT infrastructure components. The aggregated and filtered data are sent from the edge system to either an enterprise data centre or to the cloud, using cellular network, WAN and other high-speed network connectivities. The data in the cloud in a few industry-specific use cases are used to create a digital twin of the physical system, which is used for predictive analytics and simulations. The data in the cloud are managed using a CPS/IoT platform that provides data services, like de-duplication, search operations, data query services, sanitization and policy-based archival.

Advanced analytics are applied on the centralized data lake to derive new insights, which, in turn, are used to control and optimize the working of the Cyber Physical System. In many implementations, the CPS data platform also supports deployment of various industry-specific applications to meet the needs of the business and the targeted customer base. This distributed system architecture, from edge to cloud, with heterogeneous technologies, starting from different industrial bus architectures, control systems, compute infrastructure, wired and wireless communication and numerous applications, make securing any Cyber Physical System an enormous challenge.

Let us take a deeper look into the technology elements and security operation challenges of a typical Cyber Physical System, spanning OT to IT technologies. The key elements in the OT infrastructure contains sensor and actuator networks and various embedded control systems, designed using industry-specific technology and a communication bus like Supervisory Control and Data Acquisition (SCADA), Profibus, Modbus etc. These industrial bus standards and protocols are several decades old and were designed when the security threat landscape was completely different. The OT infrastructure also includes a special-purpose Programmable Logic Controller (PLC) device, that is used to monitor the sensors and control the actuators. These systems normally run very old versions of Operating Systems (OS) that, in many cases, may be out of support due to changes in component technologies and innovations. These systems are also not regularly assessed for vulnerabilities and the update to newer firmware and application stack is normally done infrequently, due to the potential risk of downtime. These issues increase the risk of cyber-attacks, especially in the modern world where more and more physical systems are connected to the Internet.

With increased focus on security, privacy and localization of data for regulation and compliance reasons, edge computing is gaining greater momentum in modern deployments. Many Cyber Physical Solutions have an additional layer of computing to perform complex analytics and inference jobs, that are used to optimize and control the data transmitted to the cloud. Data security becomes very important at both the edge and the cloud as the operations can be impacted due to tampered or unavailable data. Communication to the gateway/edge computing from the physical systems happens through wired or low-power wireless protocols like Long Range (LoRA), Zigbee Blue Tooth Low-Energy (BLE), etc. Many of the wireless protocols are subjected to spoofing and replay attacks, and wireless communication can easily be acquired and modified, using jammers and other sophisticated devices.

With PLCs, gateways and edge devices, potentially coming from different vendors with different architectures, ensuring that these devices are configured and hardened based on assessed risks and emerging threat landscape becomes a big challenge. Compounding this problem is the fact that the OT administrators may not be skilled in managing IT systems effectively. With new low-power and wireless communication technologies gaining popularity and acceptance, it becomes very important to look at good security practices and operational procedures to secure the CPS solutions. Advanced security practices and use of modern technologies is key to dealing with motivated cyber criminals and nation state-sponsored attacks.

14.3.1.3 CPS Environment and Ecosystem Challenges

A traditional data centre has a lot of physical security controls and mechanisms, such as wired fence, security guards, boom barriers, doors with access control, multifactor authentication and locked racks and cages, setup to prevent unauthorized access to the computing infrastructure. In comparison, CPS edge computers, used for collecting and controlling the devices, are installed on manufacturing floors, usually in a harsh physical environment (e.g., nonstandard temperature and humidity) with non-optimal physical security. Any intrusion on the edge server in charge of data management and control logic can result in potential compromise of the physical machines and the data that is used to derive insights, which is key for the reliable working of any CPS solution. It is very important to ensure that the systems controlling the CPS have the right level of physical security and are protected from external hackers and internal threats, including disgruntled employees.

14.3.2 Opportunities and Key Solutions for Securing Cyber Physical Systems

Let us now look at some of the key security best-practices and technologies for securing Cyber Physical Systems. Looking at the modern threat landscape and the complex and distributed designs, all Cyber Physical Systems have to be built with security in mind, taking into account the needs of the digital economy. This means that components powering the Cyber Physical Systems, including the sensor network, embedded data acquisition and control systems, edge computing devices, communication network, data centre/cloud infrastructure, CPS/IoT software platform and applications, have to be secured continuously for dealing with the complex and changing threat landscape. Let us take a look at some of the opportunities and key practices in the area of product security, secure development methodology and secure monitoring and operations of a Cyber Physical System.

14.3.2.1 Securing the CPS Components and Solutions

It is a well-known fact that security is only as strong as its weakest link. It is important to ensure that all of the components that make up a Cyber Physical System are designed with the appropriate security features, using an holistic approach starting from OT to IT technologies. Since CPS is designed with a distributed computing paradigm with many embedded systems, it is important to ensure that all the installed firmware and software components are compliant with customer policies and approved security baselines, while being free from any kind of malware. Let us look at some key security features that are important in the context of Cyber Physical Systems.

14.3.2.1.1 Hardware Root of Trust in Cyber Physical Systems

One of the mandatory requirements for a secure CPS design is the presence of malware-free/uncompromised firmware and software. One way of implementing this in critical embedded systems is using the hardware root of trust feature that allows the verification of the integrity and authenticity of the firmware and software of all key system components inside a device. As part of this feature, all embedded systems should be designed with a hardware root of trust module that verifies the digital signature of all the components inside adevice and ensures that a device boots up only with valid firmware and software. A properly implemented hardware root of trust in a silicon chip inside the embedded system can also be used to verify supply chain integrity and any compromise of the software stack at the supply chain or during transit.

14.3.2.1.2 Secure CPS Cryptographic Module Design

CPS systems will have multiple communication modules over different hardware bus architectures, that includes wired and wireless protocols. To ensure high levels of security, it is recommended that all critical communication with and controls of the device are carried out using secure communication protocols, that can guarantee confidentiality and integrity at all times. It is highly recommended that key communication between entities outside the trust boundary should support mutual authentication using certificate-based validation or other industry-accepted methods, and that all commands and data should be sent over a communication channel providing strong encryption and integrity verification. For devices with low power requirements, elliptic curve cryptography (ECC) with key strength of 256 bits and higher is a good recommended practice. For internal communications within a data centre or between two internal devices, use of Advanced Encryption Standard (AES) with a key size of 128 or greater and operating in Cipher Block Chaining (CBC) or Galios Counter mode (GCM) is considered a secure option. All communications external to a network and across trust boundaries should use secure protocols. Data from the OT to IT boundary should be sent over secure protocols like HTTPS, using TLS1.2 or greater, and all systems should support Transport Layer Security (TLS) 1.2 protocol and above, with weaker protocols/cryptographic algorithms disabled by default.

14.3.2.1.3 Trusted Computing and Attestation Solutions for Cyber Physical Systems

The Trusted Computing Group (TCG) has defined multiple specifications to verify trusted computing systems and to securely attest all computing devices, IoT systems and edge gateways. The use of Trusted Platform Module (TPM) as a hardware root of trust enabler allows measurement of the code and configuration of all the components within an embedded system or any infrastructure device, like server, storage and networking equipment. The measurements from the device can be retrieved by a centralized remote attestation server, using a secure communication channel, and the measurements can be verified using whitelisted repositories of approved versions of software and configurations. This attestation solution, using a hardware device, such as TPM, will help to detect the integrity of the software running on the device, along with any unapproved and unauthorized devices.

14.3.2.1.4 Key Product Security Features

Products used in critical CPS deployments should support strong authentication features, allowing the password length and complexity to be configured, based on the required level of security. The embedded end points should support installation of different types of certificates, which can be used for authentication, identification and secure connection to remote devices. It is also important to ensure that the products are designed with no back doors or hidden accounts/keys, as motivated attackers can easily reverse-engineer the code base to get all the secrets embedded in the product. The keys used in encryption should be protected in a trusted store like Trusted Platform Module (TPM) and the device should have protection from common Denial of Service (DoS) attacks, using secure implementation and handling of invalid input and excessive resource consumption.

All embedded systems used in a typical Cyber Physical System design have to support secure firmware/software update mechanisms. As part of this feature, the products should support updates to a newer and more secure version of the software and also ensure that only authentic and non-compromised firmware/software can be flashed to the system. This validation can be performed using digitally signed firmware/software components and the module responsible for firmware/software update should ensure digital signature validation before a firmware/software update action is initiated on the device.

14.3.2.2 Secure Development Methodology for Hardware and Software Powering the CPS

To ensure a secure implementation of Cyber Physical Systems, it is very important to design with built-in security, using the right methodology and development practices to ensure that the product meets the required security assurance level. All component vendors should understand the key standards and regulations during the requirement phase and ensure compliance during the development and testing phase. For example, product development teams should be aware of newer regulations, such as the Californian law that prohibits standard default passwords for IoT devices. The UK government is also mandating some basic security practices for all Internet-connected devices and a proposal to use security labels on the device that are compliant with the new regulations.

Another key practice that is important during any product development is the use of threat-modelling practice during the design phase. Threat modelling starts with a well-defined security architecture capturing all the key CPS components, applications, the users, the key data flows and the data stores that are part of the physical and cyber system. Using known attack patterns on key assets, a detailed analysis and review of the architecture is carried out to proactively identify design flaws and to mitigate the important issues, taking into account emerging risks and threat landscape. There are multiple threat modelling methodologies, with the most popular being the Microsoft methodology called STRIDE. STRIDE is an acronym for Spoofing, Tampering, Repudiation, Information disclosure, Denial of Service and Elevation of privilege. Performing a threat model for any CPS involves capturing all the IT and OT components, interfaces and key assets that could be attractive to cyber criminals. The threat model is then used to identify threats and vulnerabilities across the IT and OT landscape and the potential impact.

A typical threat modelling exercise will involve assessing the threats on all key IT components, including edge computing, communication network, etc., and the interconnections between the IT components and the physical systems, compromising control systems, sensor and actuator networks, and distributed embedded systems. During a CPS threat modeling exercise, it is important to assess the impact of a malfunctioning or compromised sensor, spoofed or tampered sensor data or a compromised edge computing system that can abnormally control the key actuators required for proper functioning of the Cyber Physical System. It is also important to look at the impact due to compromised applications and corrupted data stores used for analytics and deriving insights. Based on the criticality of the identified issues, appropriate mitigations, that include strengthening the confidentiality integrity, availability, authorization and non-repudiation of the system, needs to be added to the design and verified during the test phase. More details on the Microsoft threat modelling methodology can be found using the link https://www.microsoft.com/en-us/security engineering/sdl/threat modeling.

During the validation phase, it is important to ensure that the product undergoes vulnerability scanning to detect all well-known vulnerabilities in the system. It is highly recommended that all critical and high-level vulnerabilities are fixed before the release of the product. For products that have management access points connected directly to the Internet (e.g. edge computing devices, IoT gateways), it is highly recommended for the device to undergo penetration testing to detect unknown vulnerabilities that can compromise the system and devices connected to the same network. Along with the penetration testing of the individual components that make up the CPS system, it is also highly recommended to perform penetration testing of the complete Cyber Physical System solution stack during the commissioning phase to discover potential weaknesses in the configuration of firewalls, network infrastructure and host-based technical controls. A detailed list of important secure development practices is documented as part of the open Software Assurance Maturity Model (SAMM). The details of open SAMM can be found at https://www.opensamm.org/

14.3.2.3 Secure by Default and Advanced Physical Security Features

Secure by Default is an important attribute for critical Cyber Physical System components as many products generally get commissioned with minimal configuration and human effort on a factory floor. The Cyber Physical System should be secured by default at the time of shipping from the factory, with all weak protocols and insecure services turned off or disabled completely.

To handle the environment where the edge computers may be installed, on factory floors with little physical security, it is recommended that all key edge computing devices are installed with physical intrusion sensors to detect events like chassis opening actions aimed at stealing or compromising the platform. It is a recommended design that these intrusion sensors should operate using a local battery source to support intrusion detection even when the external power is switched off completely. Such advanced security controls within the device, along with CCTV cameras and real-time video analytics, can help in providing optimal security for computing devices installed in harsh environments like factory floors and smart grid generation stations. It is also highly recommended that modern CPS components should have a way to securely alert centralized management stations when a security anomaly is detected in the environment.

14.3.2.4 CPS Secure Operations

Once the appropriate set of secure products is identified and assembled to create a CPS solution, one of the main challenges is to securely operate the Cyber Physical System 24/7 while meeting all applicable compliance requirements. To meet the needs of the future, the CPS Security operations has to be adaptive, automated and analytics-ready, while meeting the business needs of agility, reliability and availability. Here are a few important considerations for securing the Cyber Physical System operations.

14.3.2.4.1 End-Point Protection

With the complexity of Cyber Physical Systems, including the interconnection between physical and cyber systems, it is important to look at security holistically, considering the potential threat landscape. With the distributed nature of Cyber Physical Systems, the network-centric security controls for perimeter security, with firewall, intrusion detection/prevention systems, etc. will no longer be sufficient, when considering either external or insider threats. With modern attackers being highly motivated and resourceful, it is important to look at security beyond the network layer, using end-point and unified security products and solutions. It is very important to install a next-generation end-point protection solution that includes host-based intrusion detection system and/or host firewalls to prevent and detect unauthorized access at the endpoints. It is also important that all critical systems used in Cyber Physical Solutions run the latest version of antimalware programmes, with the virus signatures being updated regularly to deal with newly discovered vulnerabilities. A combination of the latest antimalware, host-based intrusion detection systems and firewalls on edge computing systems, cloud systems and all embedded systems that perform critical operations, will ensure a good level of protection from many advanced persistent threats.

14.3.2.4.2 Segmentation and Network Isolation

Since IoT and CPS security is evolving and the business is always in a race to release the next innovative feature/service in this highly competitive world, sometimes there is a good chance that the Cyber Physical System deployment may contain devices with non-optimal security features. Some examples of the non-optimal features include the configuration capability to turn off weak protocols/ports, lack of features to update self-signed certificates and the lack of provision to set password complexity or enable two-factor authentication. In some cases, the design of the device may make the firmware/software update process very cumbersome or the vendor patch update policies may not meet the required security timelines. In such cases, with a mix of multiple devices, network segmentation and the compartmentalization of secure devices from devices with non-optimal security features, using properly configured firewalls, is going to be a key security operations practice for all Cyber Physical Systems. Depending on the nature of the device, available product security features and certifications and the risk caused by connecting an unsecured device to the Internet, the network has to be partitioned into different isolated areas, with insecure devices partitioned in a separate logical network behind a well-configured firewall. It is also a highly recommended best practice to separate the IT and OT network, using proper technical controls and firewall policies and mandated multifactor authentication for all administrator-level access to critical resources.

14.3.2.4.3 Hardening Cyber Physical Systems

For protection against cyber criminals and nation state-sponsored attacks, all Cyber Physical Systems should be hardened, based on the identified risks, while balancing usability, performance and flexibility. Hardening of a Cyber Physical System involves selecting the right set of technical controls for protecting the key physical assets, processes and data required for the proper operation of the solution. It is highly recommended that a security baseline is identified for all components that are part of the OT and IT stack, and the systems are configured according to the defined baselines. Some of the best practices followed during security hardening and baseline creation are the use of strong identification and authentication procedures. The solutions and operations should be built with the principle of least privilege, using proper access controls that limit the users to performing the operations required for their job level/role. It is also important to secure the data at rest and transit by maintaining the confidentiality, integrity and non-repudiation at all points of time, using data encryption, digital signature/hash of files, authorization and audit logging procedures. It is also recommended that all systems are installed with the latest security patches and the Cyber Physical Systems are audited for compliance with defined policies and applicable regulations before commissioning the solution.

With the constant threat of newer types of attacks that do not follow a well-defined pattern, analytics is increasingly becoming a key driver for mission-critical Cyber Physical Systems. It is also important to protect the data and to limit exposure to ransomware types of attacks that have taken many industries by surprise in the past few years. So, Cyber Physical Systems must comply with standard IT security practices, like ensuring regular backup on multiple media, while also ensuring that the design is fault tolerant, highly available and, in some cases, disaster recovery ready.

14.3.2.4.4 Proactive Vulnerability Management

With increased risk of zero-day attacks and with increasing numbers of breaches happening as a result of security patches missing on critical systems, proactive vulnerability assessment and remediation is a key operational practice for any Cyber Physical System. With proliferation of open-source products and operating systems, like Linux, in both enterprise and OT devices, it is very important to have a proactive vulnerability management practice which allows discovery of newly discovered vulnerabilities and assessment of their potential impact. The recommended practice is to have automated solutions for monitoring new vulnerabilities and to ensure timely triaging of these issues, based on the criticality of the vulnerability and associated risks. The National Vulnerability Database (NVD) is a good source of information on newly discovered vulnerabilities, and there are many proprietary tools, from companies like Synopsis (Blackduck tool), Flexera (Code insight software), etc., that can help in managing open-source component discovery and vulnerability management in a simple and automated way.

14.3.2.4.5 AI-Based Security Monitoring System

With the threat landscape and attack pattern becoming very complicated, signature-based detection mechanisms found in traditional anti-malware solutions may not be adequate for the modern systems. It is important to look at AI-based security analytics solutions, with user and entity behaviour analytic capabilities to detect anomalies in the environment, device or data communication layers. For IT systems, Security Incident and Event Management (SIEM), with capabilities to correlate logs from different sources, should be considered as part of the overall solution. It is also highly recommended to turn on security alert capabilities from all embedded systems and IT computing devices. A well-documented security incident management process and procedure and an AI-based automation and recommendation solution that limits false positives should be used for managing all potential security incidents from a Cyber Physical System.

14.3.2.4.6 Periodic Assessment and Audit of Cyber Physical Systems

Cyber Physical Systems are much more complicated than traditional IT systems and the operations of these interconnected IT and OT technologies can be prone to manual errors. It is very important to continuously assess that the configuration of network segmentation, assets and network security controls are in accordance with approved baselines and ensure compliance with key security requirements. To detect unauthorized devices that may have been added without proper approval or inserted deliberately by cyber criminals, it is important to periodically discover all the elements in the network and match them with a whitelisted/approved set of asset register or asset inventory database. Along with the periodic monitoring of security controls and newly added assets, it is also highly recommended to periodically perform penetration testing of the Cyber Physical System, taking into consideration the IT and OT technology usage, emerging threats and vulnerability landscape.

14.3.3 Emerging Technologies and Opportunities for Securing Cyber Physical Systems

CPS security will continue to evolve as the Industry 4.0 vision starts gaining more traction. There are many emerging technologies that open up new challenges/opportunities in the area of Cyber Physical System security. Here are a few of the main emerging technologies that can have an impact on CPS security and solution design in the near future.

14.3.3.1 Quantum Computing Safe CPS design

With Industry 4.0 gaining a lot of attention and momentum, CPS will become mainstream, with multiple deployments expected in the coming years. At the same time, there are lots of new technology trends that can have an impact on the security of CPS systems. Quantum computing advances can impact the security of IT and OT systems and, if the developments stay at the same pace, many existing encryption algorithms will become easily compromised, leading to potential compromise of the security of Cyber Physical Systems. With this trend in mind, it is advisable for modern Cyber Physical Systems to be designed with cryptographic agility, allowing the configuration of cryptographic cipher key size (e.g., key size of 3072 and above for RSA algorithms) and cipher algorithms to handle the advances in quantum computing without major changes to the product. It is highly recommended that important data at rest is protected by strong encryption algorithms and strong key sizes. Please see references for more information on quantum computing and impact on security.

14.3.3.2 Blockchain for Supply Chain and Identity Management

Blockchain is a new technology that has promise to solve many problems in the area of identity management and supply chain security. Because of its distributed nature, Blockchain can be a good option for enhancing supply chain security of Cyber Physical Systems by tracking the components during transit across all supply chain phases. There is lot of interesting work taking place in the industry in the area of Blockchain, and many research papers are available on the Internet. It is important to look at the maturity of the Blockchain-based solutions and to leverage this to secure Cyber Physical System implementations, especially around identity management and secure supply chain.

14.3.3.3 Advanced Security Hardware Accelerators

With increasing adoption of CPS and IOT technologies, it is important for the industry to look at hardware-based accelerators and security co-processors for cryptographic functions and advanced security features. Hardware-based accelerators have the advantage of low power consumption and increased security but come with a disadvantage of difficulty in upgrading, especially if a new vulnerability is found. There is also a need to look at new energy-efficient cryptographic algorithms that can be run on battery-operated backed-up devices, while requiring very little maintenance. The area of advanced security hardware accelerators that are tailored to the needs of CPS could be a good research area for the future.

14.3.3.4 Advanced Security Analytics for Cyber Physical Systems

The Cyber Physical System design is very complicated due to the interconnection between physical electromechanical machines and IT infrastructure. The CPS threat landscape is complex, with the impact of an attack being significantly greater when compared with a similar attack in the cyber space, with Stuxnet being a good example. This complex threat landscape and attack patterns create new opportunities for research activities around near real-time security analytics and advanced machine learning algorithms, that have the potential to detect newer attack patterns and anomalies with a very high levels of accuracy.

14.3.3.5 Industry-Specific Reference Threat Models and Threat Intelligence

CPS is an evolving area and the reference architecture and implementation varies from industry to industry. It is a big opportunity for researchers and standard bodies to create reference threat models, identifying top-ten threats and associated mitigations along similar lines to the Open Web Application Security Project (OWASP) Top 10 for web applications and OWASP Top 10 for IoT devices. This body of work should ideally be made available to all key stakeholders and can be created with the help of researchers, threat intelligence and key industry security experts. A widely and easily understood threat landscape, attack vectors and mitigation information can help system integrators, service providers and operators to secure Cyber Physical Systems, using standard implementations of technical controls and secure product design, which, in turn, can help the Industry in securing the critical infrastructure.

14.3.3.6 Converged Edge Computing

With IT systems being increasingly used in the edge (e.g. manufacturing floors), there is an opportunity to look at new system designs and solutions to standardize and modernize the Cyber Physical Systems. There is a new line of infrastructure products/solutions, like Hewlett Packard Enterprise's Converged Edge computing hardware that converges OT functionality, such as data acquisition and control, with traditional IT capability. These systems are designed for the harsh environments found on typical factory floors and can run unmodified versions of complex enterprise applications while supporting both wired and wireless connectivity. These new systems, with general-purpose x86 and ARM processor, also support multiple Input/Output modules to interface directly with sensor and actuator devices. The benefit of this convergence results in greater security, higher performance and lower operating expenditure, through lower space and energy requirements. Such types of modern infrastructure design can help in simplifying the Cyber Physical System solution stack, thereby helping the overall security.

14.4 Conclusions

Cyber Physical Systems are an important part of the modern digital world, which is increasingly becoming smarter, connected and automated. Cyber Physical Systems with deep connections between the physical systems and IT/Cyber technologies create new security challenges, especially when you consider the modern threat landscape with nation state-sponsored activities and advanced persistent threats. It is highly important to have an holistic view of security when designing and implementing Cyber Physical Systems, starting with well-defined security requirements and secure product designs with state-of-the-art security features. It is also important to ensure that the Cyber Physical System components are secured at the supply chain and are designed with Secure by Default methodology when shipped from the factory. Finally, during the deployment of a Cyber Physical System, we need to ensure that the solution is hardened and continuously monitored, and that proactive actions should be taken to meet the challenges of the complex and continuously changing threat landscape. Some of the technologies, design practice and security features shared in this chapter should be carefully considered when building a secure Cyber Physical System.

Links containing additional information on key topics covered in this chapter

1. https://en.wikipedia.org/wiki/Stuxnet.
2. https://ics.sans.org/media/ICS-CPPE-case-Study-2-German-Steelworks_Facility.pdf.
3. https://en.wikipedia.org/wiki/December_2015_Ukrainc_power_grid_cyberattack.
4. https://www.theinquirer.net/inquirer/news/3074922/uk-gov-iot-security-labelling.
5. https://www.microsoft.com/en-us/securityengineering/sdl/threatmodeling.
6. https://www.techwell.com/techwell-insights/2019/01/owasp-releases-latest-top-10-iot-vulnerabilities.
7. https://www.microsoft.com/en-us/microsoft-365/blog/2018/04/24/trusted-cyber-physical-systems-looks-to-protect-your-critical-infrastructure-from-modern-threats-in-the-world-of-iot/.
8. https://www.trustedcomputinggroup.org/wp-content/uploads/TCG_Guidance_for_Securing_IoT_1_0r21.pdf.
9. https://nvd.nist.gov/.
10. https://www.nist.gov/news-events/news/2019/01/nist-reveals-26-algorithms-advancing-post- quantum-crypto-semifinals.

15

Proactive UDDI for Cyber Physical Systems

Swetha Natarajan Ganeshbabu, Sridevi Sriram and Karpagam Guruvareddiyur Rangaraju

CONTENTS

Organization of the Chapter

The organization of the chapter is as follows.

Section 1 presents the terms and terminologies for the user to understand the chapter

Section 2 provides an introduction to cyber physical systems and lays emphasis on how SOA and web services can be used to overcome this problem of data availability.

Section 3 focuses on the current trends in web services and their disadvantages when used in the domain of health information systems

Section 4 conceptualizes the proactive UDDI for the cyber physical system

Section 5 illustrates the working of pUDDI with sample cases.

Section 6 discusses the validation of the designed system

Section 7 concludes this chapter.

15.1 Terms and Terminologies

Web Service: A Web Service is any logic made available to its user through Internet.

UDDI: Universal Description, Discovery and Integration is a centralised registry which holds various business related web services for public usage.

SOAP: Simple Object Access Protocol is a protocol specially designed for transmission and reception of web service messages.

WSDL: Web Service Description Language is an XML based language which describes the functional interfaces of a web service.

QoS: Quality of Service refers to behavioural quality of the web service which plays a significant role in determining then on functional aspect of any web service.

CUDA: Compute Unified Device Architecture is a platform developed specially for parallel computing by NVDIA.

15.2 Introduction

Cyber Physical Systems integrate the use of networking, computation and physical processing. One such cyber physical system is the health information system (CPS 2012). The user requirements in the field of healthcare are evolving on a day-to-day basis. The end users of every healthcare system expect more precise, reliable and accurate results within seconds. Unfortunately, the healthcare system, which is widely adopted in India, is not well connected and reliable. A patient may approach a specific hospital for diagnosis of a disease. He/she may continue treatment for several months. In such cases, the hospital will track the patient's condition and digitize the data in all ways possible. If the same person then wishes to continue the treatment elsewhere, one of the following two possibilities may occur,

1. The entire procedure of the treatment in the other hospital starts again from scratch.
2. The treatment starts at an intermediate step, with assumptions obtained from partial information available in the hands of the patient.

Either of these cases may have an adverse effect on the patient's health, and either is due to a lack of availability of the patient's health data.

The patient's health data is completely available within the original hospital which is not under the aegis of the new one. If the new hospital could access the same health record of the patient, even granular data, regarding the treatment provided, could be obtained, which would greatly aid the patient in the further treatment of the disease. This chapter aims in addressing this interoperability issue, using the concepts of SOA.

15.2.1 Role of SOA in Health Information System

SOA is a key term, describing a well-known architectural approach known as Service Oriented Architecture. SOA emphasizes applications that use the services over the network. Web services are built over the SOA, which makes it flexible, interoperable and composable. A web service is a digital piece of code, which is intended to perform a particular function over the World Wide Web (Dustdar and Schreiner 2005; Rajendran and Balasubramanie 2010; Klusch et al. 2006; Vaculin et al. 2009; Pathak et al. 2005; Maheswari and Karpagam 2015; Wang et al. 2018). A web service can be used by varied applications of diverse requirements, which makes it the best fit for a health information system.

Every hospital will have a different application to digitize the data. The application may support any type of language or may support any format of data for communication. If a different application, with support for a different data format, tries to get the data from the former application, the recipient will not understand the data that is communicated. In this case, it only ends up in the form of garbage data as there is a mismatch in the data format. To avoid this type of gibberish data being communicated, the healthcare system is in need of a service platform, which can entirely eradicate this difference that exists among the various applications. Web services play a major role in the healthcare sectors for achieving such interoperability.

15.3 Current Trends in Web Services

Every web service has three main actors (Dustdar and Schreiner 2005; Rajendran and Balasubramanie 2010; Klusch et al. 2006; Vaculin et al. 2009; Pathak et al. 2005; Maheswari and Karpagam 2015; Wang et al. 2018): web service provider, web service consumer and the registry. Web Service provider is the one which develops the web service. The web service consumer or the client is the end user of the developed web service. The web service registry is a place where the service providers publish their services and service consumers can search for the desired service to complete a specific task. The web service registry plays a very important role in a web service because, without publishing the web service in a registry, it cannot reach its intended purpose. The most commonly opted registry is known as UDDI (Universal Description, Discovery, and Integration).

UDDI (OASIS2012, Pokraev et al. 2003, Chen et al. 2003) is an XML-based registry for interoperable web services to reveal the latter's existence and visibility on the Internet. The goal is to streamline transactions by enabling service providers and consumers find one another on the Web. The UDDI registry helps in handling the visibility, code reusability, compliance, relationship, maintenance and documentation of web services, as follows:

- Achieves **visibility** by identifying services within the organization that can be reused to address a business need.
- Promotes **reusability** by preventing insignificant reinvention of codes with zero to minimal change in *modus operandi;* accelerates development time and improves productivity; categorizes the assortment of services, making it easier to manage them.
- Ensures **compliance** by using the principal of Service Oriented Architecture (SOA); dynamically discovers services stored in the UDDI registry.
- Maintains **relationships** between services, component versions and dependencies.
- Guarantees **service life cycle management** through each phase of development, from coding to deployment.

Web services can be developed for health information system in such a way that it can facilitate the information transfer across varied applications of the healthcare domain. All the web services developed for this purpose have to be registered in the UDDI registry for public usage. The main work of the UDDI registry is that, when a request reaches the registry, it discovers the relevant services and returns it to the client. The client then executes the web service to accomplish the intended task.

Every web service has a non-functional aspect associated with it (Dustdar and Schreiner 2005; Rajendran and Balasubramanie 2010; Klusch et al. 2006; Vaculin et al. 2009; Pathak et al. 2005; Maheswari and Karpagam 2015; Wang et al. 2018). The non-functional aspect of any web service describes the Quality of Service (QoS) that it offers for its client. If the QoS values are high, then it indicates that the web service is a superior web service. If not, it indicates that the particular web services lack quality.

The rate at which the web services are published in the registry is increasing at an exponential rate. The health information system is a critical application, where the failure of the information *via* a web service may lead to dangerous ill-effects on the patient. The web services that are discovered in the UDDI do not ensure the availability or the reliability of a web service. If a discovered web service from the UDDI is not available or reliable, then the data transfer becomes chaotic, and it may lead to severe health hazards for the patient. So, in order to provide superior high-quality web services for highly critical cyber physical systems, like the health information system, a strong filtering mechanism is needed in the UDDI to decide whether the services can be allowed to be published in the registry or not. In the absence of the afore-mentioned systems, the problem of UDDI becomes insurmountable in terms of services that are never preferred by the users. If this is not controlled, it will lead to slow composition of the services.

15.4 Proactive UDDI for Cyber Physical System

Proactive UDDI (pUDDI) is a UDDI-based registry which helps in filtering of low-quality web services, preventing them from being published in the registry. This decision, on accepting a web service, is based on the behaviour of the web service in terms of the non-functional aspect. Every published web service is available to be used by the client. When the client wants to use the web service provided by the web service provider, an agreement, regarding quality offered by the web service, is drawn up between the provider and the user. This agreement is called the Service Level Agreement (SLA). The SLA contains various details, regarding the quality of the web service. Most commonly found Quality of Service (QoS) parameters in the SLA are listed below:

- Response time.
- Availability.
- Reliability.

According to the agreement, the service provider assures the client that the quality of the service offered will be no less than the value provided in the SLA. The proposed system extracts these QoS values from the SLA. These extracted values are referred to as QoS_{SLA}.

QoS values specified in the SLA are verified by the registry. To achieve this verification, the registry sends 1000 Simple Object Access Protocol (SOAP) request messages. These 1000 SOAP request messages are accelerated, using GPU servers. In return, 1000 SOAP response messages are obtained. Internally, the time taken to obtain each SOAP response is recorded. Based on these, a new set of QoS values is calculated. These values are referred to as QoS_{cal}.

Based on the QoS_{SLA} and QoS_{cal} values, the acceptance of a web service for publication in the registry is decided. According to the SLA, the calculated QoS values must not be less than the QoS values specified in the SLA. If the calculated QoS values are found to be less than the specified values (a threshold of 10% variation is accepted), then it indicates that the quality of the web service to be hosted is low. Such low-quality services are rejected.

Only the web services that really satisfy the conditions given in the SLA are accepted to be hosted on the registry.

When it is decided that a web service is to be rejected, the calculated QoS values, i.e., QoS_{cal}, are copied from the database onto a spread sheet. This spread sheet is converted into a QR code. The generated QR code is then stored in the Blockchain. The service provider is informed about the rejection of this request. Figure 15.1 pictographically represents the filtering process of pUDDI.

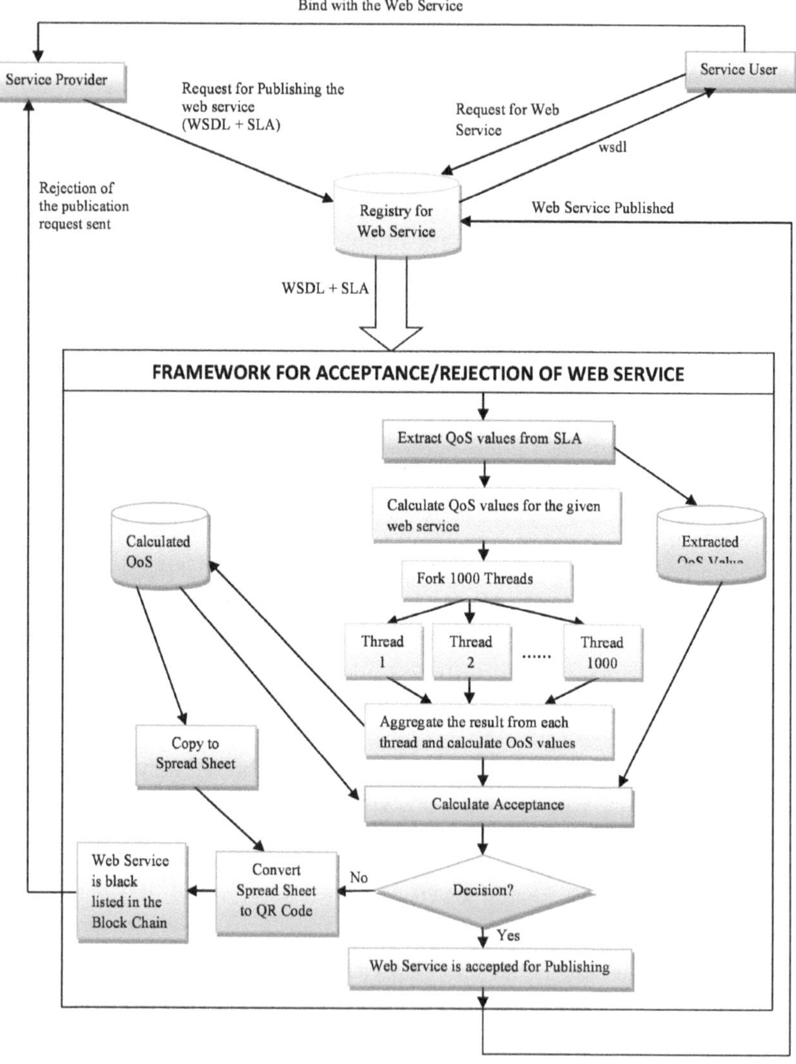

FIGURE 15.1
Framework for proactive UDDI

15.4.1 Algorithm

1. Service provider sends a request for publishing his/her web service onto the registry. The request sent contains a wsdl file along with the SLA.
2. The registry forwards the request to the framework to check if the given web service can be accepted for publishing or not.
   a. From the SLA, the specified QoS values for the web service are extracted. These values are represented as QoS_{SLA}. The different parameters of the QoS_{SLA} are listed below,
      i. $RepsonseTime_{SLA}$
      ii. $Availability_{SLA}$
      iii. $Reliability_{SLA}$
   b. With the available wsdl file, the registry tries to connect with the service provider and verifies whether the QoS values stated in the SLA are satisfied or not.
      i. To verify this, the registry sends SOAP request messages and receives SOAP response messages in turn.
      ii. This operation is performed in a parallel manner, using 1000 threads. All the 1000 SOAP requests are sent in parallel and the respective response is received.
      iii. From these 1000 response messages received, the QoS values are calculated as follows,
         1. $\text{Response Time}_{cal} = \textit{Average}(\text{Total time taken for receiving the response message})$
         2. $\text{Availability}_{cal} = \frac{\text{No. of Response messages received}}{\text{No. of Request messages sent}} * 100$
         3. $\text{Reliability}_{cal} = \frac{\text{No. of error free response messages received}}{\text{No. of Response messages received}} * 100$

   All these QoS values are collectively referred to as QoS_{cal}.
   c. Calculate the Acceptance as follows:

```
Calculate the percentage of deviation of Availability.
If percentage of negative deviation is less than 10% or no
Deviation is found:
{
  Calculate the percentage of deviation of Reliability.
  If percentage of negative deviation is less than 10% or no
  Deviation is found
  {
    Calculate the percentage of deviation.
```

```
        If percentage of negative deviation is less than 10% or
        no Deviation is found
         {
           Decision="Yes"
         }
         Else
         {
           Decision="NO"
         }
    }
    Else
    {
      Decision="NO"
    }
}
Else
{
  Decision="NO"
}
```

   d. Return the Decision.
3. If the Decision is YES, the Web Service is accepted and the service is published in the registry.
4. If the Decision is NO,
   a. The Web Service is rejected for publication in the registry.
   b. The QoS_{cal} values are copied into a spread sheet.
   c. This spreadsheet is converted into QR code.
   d. The QR code generated is added to the Blockchain and the message of rejection is sent to the service provider.

15.5 Working of pUDDI

Sample Scenario 1: A web service provider (A) has developed a web service named get Patient Details, which is used to transfer patient information from one hospital to other. The provider contacts the registry for publishing the above-mentioned service.

Table 15.1 represents the request given to the registry. A total of 1000 SOAP Request messages were sent to the provider and 1000 SOAP Response messages were received. Among the 1000 Response messages, 950 messages were error free and 50 messages were erroneous. Based on this information, QoS_{cal} values are calculated as suggested in the algorithm. Table 15.2 represents the QoS_{cal} values for Scenario 1.

TABLE 15.1

Scenario 1 Request to the pUDDI Registry

Service Provider	A
WSDL	http://ServiceProviderA.com/Webservice1/getPatientDetails?WSDL
QoS_{SLA}	$ResponseTime_{SLA}$ =10.55 ms $Availability_{SLA}$ = 98 % $Reliability_{SLA}$ = 95 %

TABLE 15.2

QoS_{cal} Values for Scenario 1

Response Time	$ResponseTime_{Thread1}$	9.55 ms
	$ResponseTime_{Thread100}$	10 ms
	$ResponseTime_{Thread500}$	9.85 ms
	$ResponseTime_{Thread750}$	10.2 ms
	$ResponseTime_{Thread1000}$	9.88 ms
Sample SOAP Request Message	<soapenv:Envelopexmlns:soapenv="http://schemas.xmlsoap.org/soap/envelope/" xmlns:pac="http://pack1/"> <soapenv:Header/> <soapenv:Body> <pac:Patientdetails> <x>C123</x> </pac:Patientdetails> </soapenv:Body> </soapenv:Envelope>	
SampleSOAP Response Message	<soapenv:Envelopexmlns:soapenv="http://schemas.xmlsoap.org/soap/envelope/" xmlns:pac="http://pack1/"> <soapenv:Header/> <soapenv:Body> <pac:CalculateResponse> <return>AAA</return> </pac:CalculateResponse> </soapenv:Body> </soapenv:Envelope>	
QoS_{cal}	$ResponseTime_{cal}$ = 10.1 ms $Availability_{cal}$ = 100 % $Reliability_{cal}$ = 95%	

Based on the QoS_{cal} values, the acceptance decision is made as follows,

- The $ResponseTime_{cal} < ResponseTime_{SLA}$. This indicates that the response was faster than that mentioned in the SLA.
- The $Availability_{cal} > Availability_{SLA}$. This indicates that the availability of the web services is also higher than the mentioned condition in the SLA.

- The $\text{Reliability}_{cal} = \text{Reliability}_{SLA}$. This indicates that the reliability of the web services is also higher than the mentioned condition in the SLA.

All the QoS_{cal} values are greater than or equal to the specified limit in the SLA. So, this web service is a superior web service and it is accepted for publishing.

Sample Scenario 2: A web service provider (B) has developed a web service named get Patient MRI Details, which is used to transfer MRI information of a patient from one hospital to another. The provider contacts the registry for publishing the above-mentioned service.

Table 15.3 represents the request given to the registry. A total of 1000 SOAP Request messages were sent to the provider and 932 SOAP Response messages were received. Among the 932 Response messages, 893 messages were error free and 39 messages were erroneous. Based on these information, QoS_{cal} values are calculated as suggested in the algorithm. Table 15.4 represents QoS_{cal} values for Scenario 2.

Based on the QoS_{cal} values, the acceptance decision is made as follows:

- The $\text{ResponseTime}_{cal} > \text{ResponseTime}_{SLA}$. This indicates that the response is slower than that mentioned in the SLA.
- The $\text{Availability}_{cal} < \text{Availability}_{SLA}$. This indicates that the availability of the web services is less than the mentioned condition in the SLA. On giving a 10% deviation, the acceptable value for $\text{Availability}_{cal}$ is 99.399 %. But the $\text{Availability}_{cal}$ exceeds the threshold of 10%.
- The $\text{Reliability}_{cal} > \text{Reliability}_{SLA}$. This indicates that the reliability of the web services is higher than the condition mentioned in the SLA.

The QoS_{cal} values are not as specified in the SLA. So, this web service is rejected. On rejection, the QR code is generated and stored in the Blockchain. Figure 15.2 represents the QR code generated for Scenario 2.

This generated QR code is published in the EthereumBlockchain (Zheng et al. 2018; Soulsby 2019; Francesco et al. 2019; Zheng et al. 2017) as open-to-public, which contains details about the blacklisted services. This system ensures tamper-proof data that can be used for statistical analysis about the web service providers.

TABLE 15.3

Scenario 2 Request to the pUDDI Registry

Service Provider	B
WSDL	http://ServiceProviderB.com/Webservice1/ getPatientMRIDetails?WSDL
QoS_{SLA}	$\text{ResponseTime}_{SLA}$ =8.50 ms $\text{Availability}_{SLA}$ = 99.5 % Reliability_{SLA} = 95 %

TABLE 15.4

QoS_{cal} Values for Scenario 2

Response Time	$ResponseTime_{Thread1}$	11.55 ms
	$ResponseTime_{Thread100}$	8.99 ms
	$ResponseTime_{Thread500}$	9.85 ms
	$ResponseTime_{Thread750}$	12.65 ms
	$ResponseTime_{Thread1000}$	12.66 ms
Sample SOAP Request Message	<soapenv:Envelopexmlns:soapenv="http://schemas.xmlsoap.org/soap/envelope/" xmlns:pac="http://pack1/"> <soapenv:Header/> <soapenv:Body> <pac:MRI> <x>C123</x> <y>2019</y> </pac:MRI> </soapenv:Body> </soapenv:Envelope>	
Sample SOAP Response Message	<soapenv:Envelopexmlns:soapenv="http://schemas.xmlsoap.org/soap/envelope/" xmlns:pac="http://pack1/"> <soapenv:Header/> <soapenv:Body> <pac:Calculate_MRIResponse> <return>MRI details</return> </pac:Calculate_MRIResponse> </soapenv:Body> </soapenv:Envelope>	
QoS_{cal}	$ResponseTime_{cal}$ = 12.65 ms	
	$Availability_{cal}$ = 93.2 %	
	$Reliability_{cal}$ = 95.81%	

15.6 System Validation

The current system of UDDI registry accepts all the requests that it receives, thereby publishing all undesirable web services into the registry. Let us assume that the current UDDI registry contains "n" services. At a time instance of t1, "m" requests are sent to the registry for web service registration. Among these m requests, p requests contain low-quality services (i.e. the QoS_{SLA}>QoS_{cal}).

If this scenario is encountered by the traditional UDDI registry:

- The total number of services after the time instance t1= n+m.
- The increase in services between the time t and t1 = m.
- The Search Space of the registry after the time instance t1 = n+m.

FIGURE 15.2
QR code generated for scenario 2

If the same scenario is encountered by the pUDDI registry:

- The total number of services after the time instance t1= n+(m−p).
- The increase in services between the time t and t1 = m−p.
- The Search Space of the registry after the time instance t1 = n+(m-p).

It is clearly seen that:

- p≥0 from the statement of the scenario
- when p=0
    - The increase in the number of services between the time t and t1 in the Traditional Registry is equal to the increase in the number of services between the time t and t1 in pUDDI.
        - m = (m−p)
        - m = (m−0)
    - m = m
    - So, the Search Space of the pUDDI is equal to the Search Space of the Traditional UDDI registry.
- When p>0
    - The increase in the number of services between the time t and t1 in the Traditional Registry is greater than the increase in the number of services between the time t and t1 in pUDDI.
        - m > (m−p) where p>0

- So, the search space of the pUDDI is less than the search space of the Traditional UDDI registry.

It is seen that the search space is either equal to that of the Traditional Registry or less than that of the Traditional Registry. But it is noted that, during the web service selection process, the number of undesirable services is higher. So, the value of p is definitely not equal to zero. So, the proposed architecture of pUDDI effectively reduces the search space by eliminating only the most undesirable services. As the search space is reduced, the time taken for selecting a web service is also reduced. Thus, the proposed architecture reduces the search time of the registry, which leads to the discovery of superior web services for cyber physical systems in a shorter amount of time than by the Traditional Registry.

15.7 Conclusion

This chapter focused on designing a pUDDI registry for filtering out the undesirable web services of lower quality. It was observed that the system developed efficiently reduces the search space of the UDDI registry, thereby reducing the execution time for the web service discovery processes. The present work has been progressed with Blockchain technology and CUDA (Compute Unified Device Architecture) programming in pUDDI, to achieve better performance aspects regarding the web service publication into the registry. As the pUDDI registry filters out the undesirable services in the publication phase, the discovered services for any crucial system (any cyber physical system) is found to be superior, thereby eliminating the need for rediscovery of services. Furthermore, this work can be extended for maintaining an efficient registry for micro services for cyber physical systems.

References

"Cyber-physical systems." *Ptolemy Project*, ptolemy.berkeley.edu/projects/cps/.

"Standards." *OASIS*, www.oasis-open.org/standards#uddiv3.0.2.

Chen, Zhou, et al. "UX-an architecture providing QoS-aware and federated support for UDDI." *Proceedings of the 2003 International Conference on Web Services*. 2003.

Dustdar, Schahram, and Wolfgang Schreiner. "A survey on web services composition." *International Journal of Web and Grid Services* 1(1) (2005): 1–30.

Francesco, Buccafurri, et al. "Ethereum transactions and smart contracts among secure identities." in *DLT@ ITASEC*, 2019: 5–16.

Klusch, Matthias, Benedikt Fries, and Katia Sycara. "Automated semantic web service discovery with OWLS-MX." *Proceedings of the Fifth International Joint Conference on Autonomous Agents and Multiagent Systems*. May 8, 2006, 915–922.

Maheswari, S., and G. R. Karpagam. "Comparative analysis of semantic web service selection methods." *Indian Journal of Science and Technology* 8(S3) (2015): 159–169.

Pathak, Jyotishman, et al. "A framework for semantic web services discovery." *Proceedings of the 7th Annual ACM International Workshop on Web Information and Data Management*. 2005.

Pokraev, Stanislav, Johan Koolwaaij, and Martin Wibbels "Extending UDDI with context-aware features based on semantic service descriptions." *ICWS*. 2003.

Rajendran, T., and P. Balasubramanie. "An optimal agent-based architecture for dynamic web service discovery with qos." *Second International Conference on Computing, Communication and Networking Technologies*. IEEE. 2010.

Soulsby, Marcus. "The benefits of the Ethereum Blockchain." *Medium Plutus.it*. 11 September 2019, medium.com/plutus-it/the-benefits-of-the-ethereum-blockchain-f332e62f7659.

Vaculin, Roman, Roman Neruda, and Katia Sycara. "The process mediation framework for semantic web services." *International Journal of Agent-Oriented Software Engineering* 3(1) (2009): 27–58.

Wang, Puwei, et al. "QoS-aware service composition using blockchain-based smart contracts." *Proceedings of the 40th International Conference on Software Engineering: Companion Proceedings*. 2018.

Zheng, Zibin, et al. "An overview of blockchain technology: Architecture, consensus, and future trends." *IEEE International Congress on Big Data (BigData Congress)*. IEEE. 2017.

Zheng, Zibin, et al. "Blockchain challenges and opportunities: A survey." *International Journal of Web and Grid Services* 14(4) (2018): 352–375.

Index